MODERN ELECTRONICS Made Simple

The Made Simple series
has been created
especially for self-education
but can equally well
be used as
an aid to group study.
However complex the subject,
the reader is taken
step by step,
clearly and methodically,
through the course. Each volume
has been prepared by experts,
taking account of
modern educational requirements,
to ensure the most
effective way of
acquiring knowledge.

D0177225

In the same series

Accounting
Acting and Stagecraft
Additional Mathematics
Administration in Business
Advertising
Anthropology
Applied Economics
Applied Mathematics
Applied Mechanics
Art Appreciation
Art of Speaking
Art of Writing
Biology
Book-keeping
British Constitution
Business and Administrative
  Organisation
Business Calculations
Business Economics
Business Statistics and Accounting
Calculus
Chemistry
Childcare
Commerce
Company Law
Company Practice
Computer Programming
Computers and Microprocessors
Cookery
Cost and Management Accounting
Data Processing
Economic History
Economic and Social Geography
Economics
Effective Communication
Electricity
Electronic Computers
Electronics
English
English Literature
Financial Management
French
Geology
German
Housing, Tenancy and Planning Law

Human Anatomy
Human Biology
Italian
Journalism
Latin
Law
Management
Marketing
Mathematics
Metalwork
Modern Biology
Modern Electronics
Modern European History
Modern Mathematics
Modern World Affairs
Money and Banking
Music
New Mathematics
Office Administration
Office Practice
Organic Chemistry
Personnel Management
Philosophy
Photography
Physical Geography
Physics
Practical Typewriting
Psychiatry
Psychology
Public Relations
Public Sector Economics
Rapid Reading
Religious Studies
Russian
Salesmanship
Secretarial Practice
Social Services
Sociology
Spanish
Statistics
Teeline Shorthand
Twentieth-Century British
  History
Typing
Woodwork

MODERN ELECTRONICS Made Simple

George H. Olsen
BSc, CEng, MIERE, MInstP

Made Simple Books
HEINEMANN : London

Printed and bound in Great Britain
by Richard Clay (The Chaucer Press), Ltd.,
Bungay, Suffolk
for the publishers William Heinemann Ltd.,
10 Upper Grosvenor Street, London, W1X 9PA

First edition, April 1977
Reprinted, April 1979
Reprinted (with amendments), September 1980
Reprinted, March 1982
Reprinted (with amendments), October 1983

British Library Cataloguing in Publication Data

Olsen, G. H.
   Modern electronics made simple.—(made simple
   books, ISSN 0265–0541)
   1. Electronic apparatus and appliances
   I. Title
   621.381   TK 7870

ISBN 0–434–98543–0

Foreword

In a technological society such as ours there can be few who do not use electronic equipment. The majority of homes now have television receivers and radios, and, increasingly, electronic apparatus of other kinds is coming to be seen as standard domestic equipment. Electronic controllers for washing machines and central-heating systems, dimmer switches for lighting control, digital clocks, tape recorders and ignition systems for cars are just a few of the many examples of how electronics affects our lives. At work the use of calculating machines, computers for banking and accountancy and the electronic control of many of our industrial manufacturing processes show just how vital is electronics to our whole economy.

This book is written for those who wish to gain an understanding of the principles of the subject. It assumes no more than a prior knowledge of elementary electricity, and, so far as possible, seeks to avoid mathematics. Although it is not intended to be a formal examination text, students taking A level GCE, TEC and City and Guilds courses in schools and technical colleges should find the book useful.

Although written in everyday language, a logical approach has been made to the subject. There has been a deliberate progression from the treatment of fundamental principles and a study of the basic devices and their characteristics, to their application in the basic circuits that are used to construct complex electronic equipment. In view of the demise of thermionic valves, these devices have not been discussed at all. It was thought to be more useful to use the space to deal with the very important semiconductor devices. After discussing the properties and behaviour of resistors, capacitors, inductors and semiconductor devices, it has been shown how these devices can be assembled into useful fundamental circuits such as amplifiers, oscillators and power supplies. The vital importance of integrated circuits has also been treated in some detail. The remaining chapters are devoted to electronic systems of general interest such as radio, television and high-fidelity sound reproduction. The increasing use of digital equipment and photocells in control and computing apparatus has also been discussed.

A chapter dealing with some of the practical aspects of the subject has been included. Ten projects based on simple, but useful, circuits have been given for those who wish to use their knowledge to produce practical results.

Questions have been included after each chapter summary. These are intended to be stimulating rather than the type set in formal examinations. Not all the answers are to be found in this book. It is the intention to stimulate a search for further knowledge in wider reading.

Finally, the author's thanks are due first to his wife who has rendered wonderful support in reading and criticizing the manuscript, and secondly to his typist, Mrs Ruth Sanderson, who has so patiently borne the indifferent handwriting and many manuscript alterations, as well as making a real contribution to completing the book on time.

GEORGE H. OLSEN

Contents

FOREWORD   v

1 INTRODUCTION   1

2 ELECTRONIC SYSTEMS, SIGNAL WAVEFORMS AND PASSIVE COMPONENTS   4
   Electronic Systems   4
   Waveforms   4
   Electronic Components   9
   Passive Components (Resistors, Capacitors, Inductors)   9
   The Response of Circuits Containing Passive Components   33
   Summary   38
   Questions   39
   Suggested Further Reading   39

3 SEMICONDUCTOR DEVICES   40
   Conductors, Insulators and Semiconductors   41
   The *pn* Junction Rectifier   46
   Point-Contact Diodes   51
   Zener Diodes   52
   The Junction Transistor   55
   Transistor Operation   57
   Silicon Temperature Sensors   66
   The Semiconductor Gas Sensor   67
   Summary   70
   Questions   71
   Suggested Further Reading   71

4 AUDIO VOLTAGE AMPLIFIERS   72
   The Basic Voltage Amplifier   74
   Biasing Circuits   75
   Graphical Techniques   78
   Multistage Amplifiers   80
   Frequency Response and the Decibel Scale   81
   Negative-Feedback Amplifiers   82
   Input and Output Impedances   88
   Frequency Selective Feedback   91
   Summary   94
   Questions   94
   Suggested Further Reading   95

5 INTEGRATED CIRCUITS   96
   The Silicon Integrated Circuit   97
   The Emitter-Coupled Circuit   101

Integrated Circuit Amplifiers   103
Summary   109
Questions   109
Suggested Further Reading   109

6 POWER AMPLIFIERS   110
   Load Matching   110
   Push–Pull Output Stages   112
   Push–Pull Driver Stages (Phase-Splitters)   117
   Summary   119
   Questions   120

7 POWER SUPPLIES   121
   Batteries   121
   Solar Cells   124
   Power from the Mains   126
   The Thyristor   137
   Control of A.C. Supplies   142
   The Triac   146
   Radio-Frequency Interference   148
   Summary   150
   Questions   151
   Suggested Further Reading   151

8 OSCILLATORS   152
   Sine-Wave Generators (Harmonic Oscillators)   153
   The Tank Circuit (*LC* and Crystal Oscillators)   154
   RC Oscillators   161
   Relaxation Oscillators   167
   Summary   170
   Questions   171

9 RADIO TRANSMISSION AND RECEPTION   172
   The Launching of Radio Waves   173
   Amplitude Modulation and Detection   175
   Sidebands   177
   Radio Receivers for Amplitude-Modulated (A.M.) Waves   181
   Tuned R.F. Amplifiers   184
   The Superheterodyne Receiver   187
   Frequency-Modulated (F.M.) Receivers   192
   F.M. Detectors   194
   Summary   201
   Questions   202
   Suggested Further Reading   202

10 THE CATHODE-RAY OSCILLOSCOPE   203
    The Cathode-Ray Tube   203
    The Oscilloscope   207

The Timebase 207
Y-Amplifiers 210
Practical Considerations 210
Voltage Measurements 211
Frequency Measurements 213
Phase Difference Measurements 214
Waveform Analysis 214
Summary 215
Questions 216
Suggested Further Reading 216

11 TELEVISION 217
The Camera Tube 219
Scanning and Synchronization 222
The Television Receiver 225
Colour Television 227
The Colour Camera 229
Summary 231
Questions 231
Suggested Further Reading 231

12 HIGH-FIDELITY SOUND REPRODUCTION 232
Characteristics of the Ear 232
High-Fidelity Equipment 234
Microphones 234
Disc Records, Pick-ups and Turntables 238
Stereo Pick-ups 241
Tape Recorders 244
High-Fidelity Amplifiers 249
Tape Recorder Amplifiers 249
Tone Controls 253
Loudspeakers 254
F.M. Stereo Radio 256
Summary 259
Questions 261
Suggested Further Reading 261

13 INTRODUCTION TO DIGITAL CIRCUITS 262
Boolean Algebra 262
The Transistor as a Switch 264
The Bistable Circuit (Flip-Flop) 268
Counting 271
BCD Numbers 271
Indicating Devices 272
Digital Computers 274
Summary 276
Questions 277
Suggested Further Reading 277

14 PHOTOELECTRIC DEVICES   278
    Photoemissive Cells   278
    Photoconductive Cells, or Light-Dependent Resistors   280
    Photovoltaic Cells   282
    The Phototransistor   284
    Summary   285
    Questions   285

15 PROJECTS   286
    A Simple Radio Receiver   288
    A Variable-Gain Amplifier   288
    An Integrated-Circuit Radio   289
    A Power Amplifier   290
    A Mains Power Supply   291
    A Stabilized Power Supply for Operational Amplifiers   294
    An Electronic D.C. Multimeter   294
    A Dimmer Circuit for Mains Lighting   296
    An Inexpensive Burglar Alarm   296
    A Light-activated Switch   297
    Further Information   300

INDEX   303

MODERN ELECTRONICS MADE SIMPLE

1

INTRODUCTION

Man has for centuries been intrigued by the nature of the universe, and more particularly the world around him. In early times it was thought that all matter was made up of four basic elements: earth, air, fire and water. Modern scientific investigations reveal a very much more complex picture. The notion that all matter is composed of indivisible finite particles or atoms is of such ancient origin that it cannot now be determined just when it arose. Sir Isaac Newton himself thought of atoms as 'solid, massy, hard, impenetrable, movable particles . . . even so hard, as never to wear or break in pieces'. It is amazing to realize that during all the centuries of pondering the nature of matter, it is only in the last seventy or eighty years that we have started to unravel the mysteries of atomic structure.

Modern theories on the structure of the atom can be said to date from the discovery of the electron in 1897 by J. J. Thomson, who was a professor of physics at Cambridge University. His work and that of Rutherford and his co-workers led to a very different picture of atomic structure from that held by Newton. According to simple theory, an **atom** consists of a central **nucleus** made up of **protons**, having a positive charge, and **neutrons**, having no charge; tiny negatively charged particles, called **electrons**, revolve around the nucleus in fixed orbits just as the planets revolve around the sun. The scattering experiments of Rutherford showed that the volume of an atom consists mostly of empty space, as does the solar system. Later workers showed that practically the entire weight * of the atom is concentrated in the nucleus. Since an atom normally is electrically neutral it will be realized that the negative charges on the orbiting electrons are balanced by an equal number of positive charges associated with the protons in the nucleus. The orbits of the electrons are arranged in 'shells' about the nucleus, each shell having a definite maximum capacity of electrons. It is the outermost shell which determines the chemical and principal physical characteristics of an atom.

In most atoms it is quite easy to dislodge electrons from the outer shell. Such electrons are known as **free electrons**. The remainder of the atom is called an **ion**. The loss of electrons results in an ion which is **positively charged**. Some atoms are able to absorb one or more free electrons that may be in the vicinity. Such atoms are then known as **negative ions** because they are negatively charged as a result of absorbing negatively charged electrons.

Electrons can exist by themselves outside of atoms. They may travel within crystals (of, say, copper or silicon) or may travel in a vacuum. Their movements will be random unless they come under the influence of an electric or magnetic field. (The term 'field' in this context means a region in which electric

* Strictly speaking the correct term to use here is 'mass'.

particles experience a force.) For example, when the terminals of an electric battery are connected by a piece of copper wire, an electric field is set up within the wire and the electrons are attracted to the positive terminal. The movement of the electrons along the wire constitutes an **electric current**. Although this orderly drifting motion is rather slow, as soon as the connection is made to the battery the impulse that initiates the flow is transmitted along the wire at almost the speed of light—that is to say, the electric field is established along the wire at almost the speed of light. The electrons are then urged in one direction under the influence of the field. A magnetic field that is constantly changing or moving can also sweep electrons along a wire. The dynamo depends upon this fact.

Electronics is the science and technology of controlling the flow of electrons to produce useful results. It is unlikely that anyone today would deny that electronics has had a tremendous impact on our lives. Television and radio receivers, hi-fi systems, computers, automatic controllers and guiding systems for rockets to the moon all depend upon the fascinating subject of electronics —and the majority of the most spectacular advances have been made only within the last thirty years.

Electronics is really a means to an end. Electronic apparatus is used to link **input** devices (called **transducers**) to **output** devices. For example, the groove on a gramophone record contains the programme material which we wish to hear. The groove is forced into a wavy pattern during the recording process. When we wish to recover the programme information it is of little use putting the gramophone record to our ear. Before the era of electronics the wavy pattern, i.e. the groove modulations, were tracked by a needle and the small mechanical vibrations were converted into variations of air pressure (which we experience as sound) by means of a diaphragm and a trumpet-shaped horn. The sounds that were heard, although recognizable, were a far cry from the original programme material. Nowadays the modulations in the groove can be converted into electrical signals by using a pick-up. This pick-up is called a transducer because it transforms the mechanical vibrations into electrical signals.

Any device that converts a physical quantity (such as force, vibration, sound or light intensity) into a corresponding electrical signal is called a transducer. Common examples are photocells, microphones, television camera picture tubes and strain gauges. Rarely is sufficient output available from the transducer to activate an output device directly. For example, a pick-up connected directly to a loudspeaker cannot produce a satisfactory reproduction of the programme material because the power output of the pick-up is far too small. Nor can the output from a television camera activate the picture tube in a domestic television receiver directly because of the distances involved. Some intermediate electronic apparatus is required. The vast majority of electronic systems follow the basic pattern shown in Fig. 1.

A good deal of electronic equipment looks complex at first sight, but it should be realized that all electronic circuits, no matter how complex, can be regarded as a combination of a comparatively small number of basic units each performing a specific function. These basic units are our 'building bricks' and from a small number of different bricks we can build many different 'buildings'. The aim of this book is to give the reader a general understanding of how individual 'bricks' work. He will then be able to understand how complete electronic systems function.

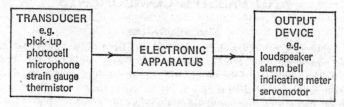

Fig. 1. The fundamental electronic system.

There is no doubt that practical work is fascinating for those interested in electronics. No amount of reading alone will make a person competent to build electronic equipment. The reader should therefore take every opportunity to do practical work. He could well buy a kit and build transistor radios, audio amplifiers, burglar alarms and so on. In the last chapter, details of simple projects are given. The projects can easily be carried out at home for a modest outlay. On gaining experience, the keen reader can then expand his activities into more ambitious experiments.

ELECTRONIC SYSTEMS, SIGNAL WAVEFORMS AND PASSIVE COMPONENTS

Electronic Systems

Fig. 2 shows the prototype electronic 'building' from which any type of electronic equipment can be built. A hi-fi system may use all of the 'bricks' shown. The amplifiers of the system will require resistors, capacitors, diodes, transistors, integrated circuits and power supplies; the radio will require, in addition, oscillators, tuners and digital decoders for stereo reception; the TV and tape-recorder section will require all of these fundamental sub-units. Some items of electronic equipment may need fewer sub-units. An oscillator, for example, requires only resistors, capacitors (and possibly inductors), as well as transistors, diodes and power supplies. An oscillator is a piece of apparatus that produces regularly changing voltages. Examples of these changing voltages are given a little later in the chapter when we discuss waveforms. Oscillators are useful for testing amplifiers, and are themselves essential sub-units of radio transmitters and receivers. They are also used in digital computing and control circuitry.

It will be seen that the whole of modern electronic circuitry rests on the foundation of resistors, capacitors and inductors as well as the semiconductor devices, viz. diodes and transistors. It will be necessary for the reader to understand the way in which such components work before he can understand how electronic circuits work.

Waveforms

An examination of a piece of electronic equipment under working conditions will reveal many different voltages and currents at various points throughout the circuit. The useful signal information (which is obtained from aerials or transducers) is carried by voltages and currents that vary, so we frequently consider graphs which show how the size or magnitude of these electrical quantities alters during a given time. The shape of such a graph is known as the **waveform** of the voltage or current being studied. In many cases the waveforms are **periodic** (i.e. they are repeated exactly in equal successive intervals of time) so the resulting graphs have a regular pattern.

Experience in interpreting the waveforms in an electronic circuit often produces valuable information about the behaviour of the electronic apparatus or associated equipment. One of the most useful pieces of electronic apparatus for examining waveforms is the **cathode-ray oscilloscope** because the graphs are automatically displayed on a screen. So useful is this item of test equipment that we shall discuss the way in which it works in some detail at a later stage in the book.

It is important for the beginner to understand what is meant by the terms 'voltage', a 'rise or fall of voltage in circuit' and 'voltage drop'. The **voltage—** or **potential difference** between two points in a circuit—is a measure of the

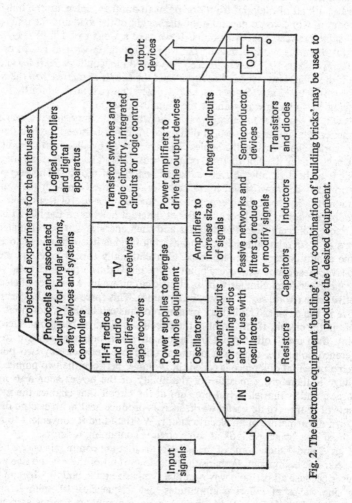

Fig. 2. The electronic equipment 'building'. Any combination of 'building bricks' may be used to produce the desired equipment.

amount of work done in moving electric charge from one point to the other at a higher potential. Let us appeal to an analogy in an attempt to clarify the situation. If we were on a hillside a certain amount of work would be necessary to raise a load of material from one point to another higher up the hill. The amount of work would depend upon the weight of the load and the difference in height between the two points. If we take a weight of 1 unit (say 1 lb. weight) then the work in raising the weight through a vertical height of say 10 feet would be 10 ft.lb. For 20 feet the work done would be 20 ft.lb. and so on. If the weight involved is 1 newton * and the height 15 metres then the work done would be 15 newton-metres, i.e. 15 joules. (There is no need for those not wishing to do so to worry about the units involved.)

It will be seen that for a unit weight the work done in raising the weight is numerically equal to the vertical difference in height between the two points. The work done in raising the bodies can be recovered by allowing the bodies to fall back down the hill to their original positions. The work done may be evident as heat or alternatively as kinetic energy of motion. The energy of motion can be transformed into electricity—as is the case with the water's motion in a hydroelectric generator—or can do other useful work, e.g. turn grinding stones to grind corn. Bodies at different heights on the hill all have, therefore, different potential energies and such energies can be calculated in terms of their relative heights. In measuring heights it is convenient to have some arbitrary zero reference level. Conventionally this is taken as mean-sea-level. Associated with every point on the hill is thus a potential relative to our mean-sea-level. Similarly, in an electric circuit we have a source of **electromotive force** (from, say, a battery or dynamo). This force can make electrons move in the circuit. The battery or dynamo thus establishes a potential gradient in the circuit. If any point in the circuit is defined as being our reference point, then every other point in the circuit has a potential relative to the reference point, i.e. there is a potential difference between any two points. Such a potential difference is called the 'voltage' between the two points. The voltage is therefore a measure of the ability of the power source to move charges, and is that electrical pressure in the circuit that enables the movements of charges to do useful work, such as produce heat in an electric fire or to turn the armature of an electric motor. We shall find it convenient to refer to these ideas when we look at some circuits containing resistors.

Figs. 3 and 4 show some of the waveforms that are commonly encountered in electronic equipment. Perhaps the most important of these is the **sine wave**. This is because all periodic waves can be synthesized, i.e. built up, by combining sine waves of differing amplitudes and frequencies. 'Amplitude', 'frequency' and 'period' are three of the most commonly used terms associated with sine waves. A **sinusoidal voltage** is one which varies with a sine wave pattern. The voltage exists across two points in the circuit. If one of the points

---

* Remember that weight is a force! 1 kg weight at the surface of the earth is the gravitational force exerted on a mass of 1 kg. At the surface of this planet this turns out to be 9·8 N; a force of 1 N is the gravitational force on a mass of just over 100 grams.

is maintained at a fixed reference potential—say by connecting it to earth (which is defined as being at zero potential)—then the potential of the other point is continually changing, first in a positive sense and then in a negative sense relative to the earth potential. The maximum difference in potential (i.e. the maximum voltage between the two points) is known as the **amplitude** of the voltage. It takes a certain time for the voltage to change from zero to the maximum positive value, then back through zero to the maximum negative value, and then back to zero again. This time is known as the **periodic time** (*T*).

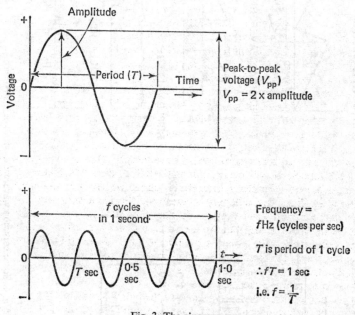

Fig. 3. The sine wave.

When the voltage changes through its various values during one period it is said to complete one cycle. The number of cycles completed during 1 second is called the **frequency** of the waveform. Thus if a voltage undergoes fifty complete cycles in a second (i.e. the sinusoidal pattern is repeated fifty times per second) the frequency, *f*, is said to be 50 Hz (Hz = Hertz, i.e. cycles per second). One cycle is completed in $\frac{1}{50}$ of second. The periodic time of the wave is therefore $\frac{1}{50}$ second, i.e. 20 milliseconds (1 second = 1,000 milliseconds). In general, $f = 1/T$.

The signals found in electronic equipment are not often sinusoidal, but they are frequently **periodic**, i.e. they have a regular repetitive pattern. The value of studying the response of circuits to sine wave signals lies in the fact, as has already been explained, that all periodic waveforms can be considered as a combination of sine waves.

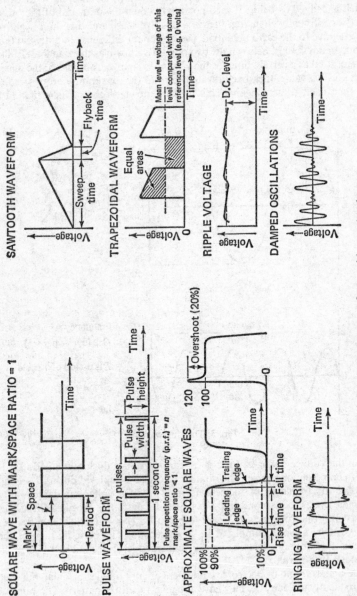

Fig. 4. Some common waveforms with associated terms.

The **square wave** of Fig. 4 is of particular importance in logic and computer work, whilst other waveforms will be seen to be important in connection with our discussions on oscillators, cathode-ray oscilloscopes, power supplies and audio amplifiers. These waveforms are shown here not only to introduce the reader to their shapes, but also to define various terms and expressions associated with them.

Electronic Components

Broadly speaking, the electronic components that are combined to form complete equipment can be divided into **active** and **passive** devices. Transistors and allied semiconductor devices are classified as active because they modify the power supplied to them. The sources of power in electronic circuits are batteries, or more conveniently electric generators that feed power into the public electricity mains supply grid. The ultimate source of power is the sun or radioactive materials; nowadays it is possible to utilize power from the sun by converting the radiant energy directly into electrical energy with the aid of solar cells. The source of power for our power stations is also the sun, such power having been stored in fossil fuels such as coal and oil. An alternative source of power for our power stations is nuclear energy.

The electric power supplied to electronic equipment can be modified to produce electrical oscillations or to amplify small signals. Although transistors are the agents by which such modifications are made, they do not actually supply power to the circuit. In spite of this, we often find it convenient to regard such devices as doing so, whilst bearing in mind that they are in fact energized from a power supply. For this reason transistors are called 'active devices'. Associated with transistors are those components that consume power or otherwise control the flow of energy in some way. These components (resistors, capacitors and inductors) are regarded as passive elements.

Passive Components

Resistors

Electric current may be regarded as a flow of electrons. The flow may be along a wire, a carbon rod, through a gas, a vacuum or through a crystal. Each electron carries a small negative charge ($1\cdot602 \times 10^{-19}$ coulombs) and when many millions of them flow along a wire a charge q will pass a particular point during a given time t. The rate at which the charge is passing (q/t coulombs per second) is called the **current**; 1 coulomb per second is known as an **ampere** (usually abbreviated to amp).

Materials vary enormously in their ability to allow the passage of a current when an electrical pressure (i.e. voltage) is applied. Silver and copper, for example, present little opposition to the flow of current. They are said to have a small resistance to current flow and are therefore known as **conductors**. Short, thick wires will pass currents of several amperes when the voltage between the ends of the wire is only a fraction of a volt. Mica, quartz, polythene and porcelain, however, pass practically no current even when the applied voltage is high. Such materials are known as **insulators**. Intermediate between conductors and insulators is the class of materials known as **semi-conductors**. Many compounds, e.g. the oxides of copper, selenium and cadmium, are known to be semiconductors; the most important semiconductor elements are germanium and silicon because transistors are made from these materials.

When a material is made into a rod or wire to form an electronic component, it is generally true to say that the current passing through the component depends upon the applied voltage and the temperature. If, for a given temperature, the current is directly proportional to the applied voltage, the component is said to obey Ohm's Law. Such components are called **resistors**. The graph of current against voltage is a straight line. A component that does not meet this requirement is termed a **non-linear** resistor, the best known example of which is the so-called thermistor. We shall be describing this component and its uses a little later in this chapter.

Frequent reference to graphs will be made when studying electronic components since pictorial representation of the data makes easier the understanding of the behaviour of both passive and active components. The graph of current through a device against the applied voltage is often the most useful one to consider; such graphs are called **characteristic curves**.

The characteristic curve for a resistor is particularly simple and is shown in Fig. 5. From the graph we note that the current is proportional to voltage, i.e.

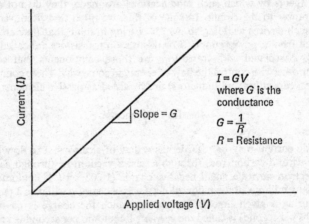

$$I = GV$$
where G is the conductance
$$G = \frac{1}{R}$$
R = Resistance

Slope = G

Current (I) — vertical axis

Applied voltage (V) — horizontal axis

Fig. 5. Graph of current against voltage for a linear resistor.

whenever the voltage is doubled the current is doubled. The ratio of voltage to current is always the same; it is constant. The constant ratio is known as the **resistance** of the component (R) and is measured in ohms when V is in volts and I is in amperes. Equally we may say that I/V is constant. This constant is often given the symbol G and is known as the **conductance**. Conductance is measured in siemens when I is in amps and V is in volts.

Resistors are used in electronic circuits to provide specific paths for electric currents and to serve as circuit elements that limit the current to some desired value. They provide a means of producing voltages as, for example, in a voltage amplifier. Here variations of transistor currents produce varying voltages across a resistor placed in series with the transistor. Resistors are also used to build networks and filters. The many uses of resistors will become increasingly apparent as the reader progresses in his study of electronics.

Fixed Resistors

Fig. 6(a) shows the construction and circuit symbols of fixed-value resistors, i.e. resistors whose value is not mechanically adjustable. General-purpose resistors are almost invariably made of a carbon composition. They are inexpensive and perform reasonably well in electronic circuits where the design requirements are not too critical. The carbon composition resistor is manufactured by hot-moulding a carbon-inert filler composition. Resistance values in the range 10 ohms to 100 megohms are readily available with power ratings from $\frac{1}{4}$ W to 2 W.

The physical size determines the power rating of the component. If this power rating is exceeded the resistor overheats and is destroyed or performs in an unreliable way. Work is done when heat is generated in a resistor. The work done, W, is equal to the potential difference across the ends of the resistor (i.e. the voltage, V) times the charge forced through the component. Suppose a current of I amperes flows for a time t seconds in a resistor which has a voltage V across it: then the charge that flows is It. This means that the work done is given by $W = VIt$. If V is in volts, I in amperes and t in seconds, the work done is in joules. If we divide both sides of the equation by t, we obtain $W/t = VI$. Now W/t is in joules per second; it is the rate at which work is being done. This is known as the **power**, P. With the given units P is in watts. Thus P (watts) = volts × amps. From Ohm's Law, $V = IR$; therefore

$$P = VI = I^2 R = V^2/R$$

Resistance values for carbon compositions are not marked on the resistor in figures, but are indicated by means of coloured rings. It is then much easier to check the values when the resistors are connected into the circuit. The rings are coloured according to a standard **colour code**. Each colour represents a digit as follows:

Black	0	Blue	6
Brown	1	Violet	7
Red	2	Grey	8
Orange	3	White	9
Yellow	4	Gold	$\pm$ 5% tolerance
Green	5	Silver	$\pm$10% tolerance

The coloured bands are painted on to the body of the resistor towards one end (see Fig. 6). The first digit of the resistance value is indicated by the colour of the band or ring nearest the end; quite often this first band is somewhat wider than the others. The next digit is indicated by the colour of the next band. The colour of the third band indicates the power of the decimal multiplier, i.e. the number of noughts (for example, orange represents 10^3—that is, three noughts placed after the preceding digits). If the third ring is gold the multiplier is 10^{-1}; if it is silver it is 10^{-2}. For example, if the first three bands are orange,

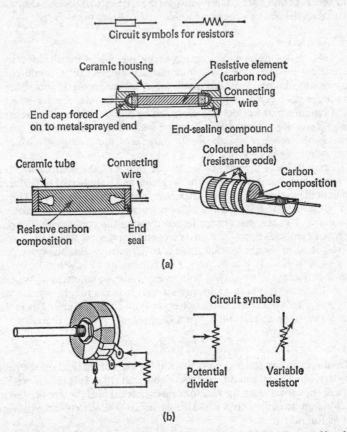

Fig. 6. Physical appearance of fixed and variable resistors. (a) Two forms of insulated carbon composition resistors. (b) The potential divider: a variable resistor is formed by connecting the slider tag to one of the tags making contact with the end of the track.

orange and gold the resistance value is 33×10^{-1}, i.e. 3·3 ohms. If the fourth band is absent the actual resistance may be within ±20 per cent of the nominal or indicated resistance. If the fourth band is silver or gold then the tolerance is ±10 or ±5 per cent respectively.

The so-called 'preferred value' system for resistance values needs explaining because of the seemingly odd numbers used. Before the Second World War the main standard values were 10, 25 and 50 with their multiples of 10. The manufacture of resistors to cover the intermediate values resulted in the production of many more resistance values than was necessary. A further disadvantage arose in the possibility of finding resistors of given nominal markings with resistances greater than some components having a higher nominal marking. The overlapping is illustrated by Fig. 7. The old system shows the overlapping

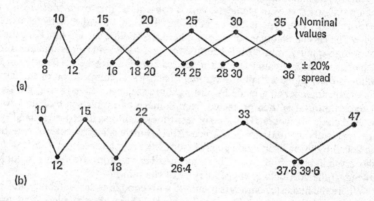

(a)

(b)

Fig. 7. (a) shows the degree of overlapping with the old system whilst (b) shows the tolerance spreads using the preferred value system (upper figures give the nominal value). The tolerance in each case is ±20 per cent.

that is now avoided by manufacturers. The most efficient system is based on a logarithmic scale. It can be shown that for a tolerance of ±20 per cent the minimum overlapping occurs when the nominal values are 10, 15, 22, 33, 47, 68. For the ±10 per cent range the figures are 10, 12, 15, 18, 22, 27, 33, 39, 47, 56, 68 and 82.

In specifying the values of resistance up to 999 ohms the number of ohms is used (e.g. 470 Ω). For resistance values between 1,000 and 999,999 the number of thousands (i.e. kilohms) followed by k is used (e.g. 680k, 1·2k). From one million—i.e. one megohm—upwards the number of megohms is used followed by M (e.g. 4·7 M). (In electronic texts the symbol Ω for ohms is frequently omitted.) To overcome the ambiguity that may arise when printing quality is poor or reproductions of circuit diagrams are made with automatic reproducing equipment, the decimal point is now frequently replaced by the multiplier symbol. Thus 2·7 kΩ is printed as 2k7 and 1·5 MΩ as 1M5. Spurious dots on the page cannot then be confused with decimal points.

Circuit designers frequently find that carbon composition resistors suffer from several disadvantages. Among the most important of these disadvantages is the comparatively poor stability of resistance value with changes of temperature and ageing. For work with very small signal voltages the 'noise' introduced by composition types can be troublesome. 'Noise' is the production of small unwanted voltages as a result of thermal agitation within the resistor. The term 'noise' arises because of the hissing produced in a loudspeaker when audio equipment is involved. The noise voltages are produced over the whole range of audio frequencies. Readers will be aware that if the hissing is loud compared with the wanted programme material, unpleasant reproduction results; the ratio of signal-to-noise should be as high as possible. Even when audio equipment is not involved it is still convenient to refer to the random-frequency unwanted voltages as 'noise'.

Improvements in manufacturing techniques have led to two other types of commonly used resistor elements, with a consequent improvement in resistor performance: one type is the carbon film element, the other a metal oxide type. The **carbon film type** consists of a hard, crystalline carbon film deposited on a high-quality, non-porous ceramic former. A nickel layer is plated on to the ends of the body and terminating wires are then attached. The resistance value is determined initially by varying the chemical composition of the film, final adjustments being made by cutting a precise, clean-edged spiral groove. The carbon film is then protected from damage and moisture by baking on several coats of lacquer. This type of resistor element has improved reliability, thermal stability and noise characteristics. Further improvements can be achieved by using a tin oxide glassy layer as the resistance element instead of the crystalline carbon film. These **metal-oxide** resistors are highly reliable owing to the particularly rugged nature of the film.

Where the highest accuracy is required, with good long-term stability, precision types of **wire-wound** resistors are used. These are the types that are incorporated in high-grade multimeters. Very close tolerances are possible, a figure of $\pm 0 \cdot 1$ per cent being common. Wire (usually nichrome or manganin) is carefully selected and wound onto bobbins or sectionalized spools. After artificial ageing and stabilizing, the resistors are vacuum-impregnated with varnish or some other suitable sealing material.

Where high-power dissipation is encountered (say from 3 W upwards), general-purpose wire-wound resistors are used. The wire is wound on to ceramic formers and the whole assembly is protected by cement, lacquer or vitreous-enamelled coatings. If the wattage dissipation is really high (say 50 W and above), it may be necessary to adopt an open-wound construction and protect the winding with a metallic grill or housing. Since this type of resistor is used for purposes not requiring high precision, 10 or 20 per cent tolerances are usual.

Variable Resistors

For many purposes it is necessary to be able to alter at will the resistance value in the circuit, e.g. to control the volume of sound from a record player or to alter the brightness of a TV picture. For these purposes a general purpose 'volume-control' is used. This type of variable resistor is often called a potentiometer although the use of this term is incorrect since actual measurement of potential is rarely involved. A better term to describe this component is **'potential divider'**. Electronic engineers and amateurs often use the jargon term 'pot' for a potential divider. The general-purpose variable resistor has a carbon track produced by spraying carbon suspension on to a plastic strip and curing the suspension at high temperature. Rectangular strips are then formed into the arc of a circle subtending an angle of about 300°. Alternatively, the track may be on a circular disc. Electrical connections are made to the two ends of the resistive track and brought out to solder tags. A mechanically adjustable wiper arm makes contact with the track. The connection to the wiper arm is brought out to a third tag. By moving the position of the arm it is possible to select a suitable fraction of the total resistance. For potential division the total voltage is applied across the resistive element. A variable fraction of this voltage is then available between the wiper arm contact and either of the remaining tags.

Cermet trimmers and potentiometers are becoming increasingly popular now that large-volume sales are reducing the price. The cermet element uses a high-purity alumina substrate upon which is 'fired' a metal-oxide film. The name 'cermet' arises because of the use of a ceramic base with a metal-oxide film. Compared with sprayed carbon types the elements are at least ten times more stable, have superior mechanical life and can be rated at many times the wattage rating for the same size.

Various so-called 'laws' of variable resistors are available. A linear potential divider is one in which the resistive track or wire is uniform throughout its length. Equal changes of angular rotation of the shaft bring about equal changes of resistance between the slider terminal and one end of the track. Several other laws are available, the choice of which depends upon the application. In volume controls it is desirable that equal changes in shaft rotation bring about equal changes in loudness. This is achieved by using a logarithmic law for the track. Other laws such as semi-log, inverse log, linear tapered, etc., are available for special applications. In some scientific applications it is convenient to have sine-wave and cosine-wave laws.

It is possible to have two potential dividers ganged together so that a single spindle operates both wiper arms simultaneously. A frequent application is in stereo work, where two separate amplifier stages are used. The stereo signals can be processed by each amplifier separately, but it is a great convenience to have the volume of sound produced by each amplifier controlled by a single spindle.

Series and Parallel Arrangements

Resistors are said to be **in series** when they are connected together to form a chain. If the ends of the chain are connected to a source of electric power, e.g. a battery, then the current has the same value or magnitude at any point in the chain. Across any given resistor a voltage exists. It is the voltage (or potential difference) between the ends of a resistor that causes the current to flow. The size of the voltage is calculated from Ohm's Law, viz. $V = IR$. If V_1, V_2 and V_3 are the respective voltages across three resistors R_1, R_2 and R_3, as in Fig. 8(a), then the supply voltage V is found by adding $V_1 + V_2 + V_3$. Since

$$V = V_1 + V_2 + V_3 = IR_1 + IR_2 + IR_3 = I(R_1 + R_2 + R_3)$$

we see that the total resistance in the circuit is the sum of the individual resistances.

Resistors are said to be **in parallel** when each resistor is connected across the power supply as in Fig. 8(b). Since the voltage across each resistor is the

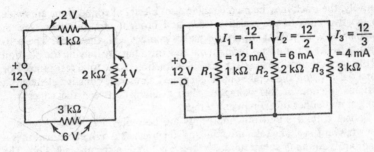

(a) Total resistance
$= 1 + 2 + 3 = 6 \text{ k}\Omega$

Current in circuit
$= \dfrac{12 \text{ V}}{6 \text{ k}\Omega} = 2 \text{ mA}$

(b) Total resistance R given by
$$\frac{1}{R} = \frac{1}{1} + \frac{1}{2} = \frac{1}{3} = \frac{6+3+2}{6}$$
$$= \frac{11}{6} \text{ millisiemens}$$
$$\therefore R = \frac{6}{11} \text{ k}\Omega$$
$$\text{Total current} = \frac{12 \text{ V}}{6/11 \text{ k}\Omega} = \frac{12}{6} \times 11$$
$$= 22 \text{ mA}$$

Fig. 8. Series and parallel arrangement of resistors.

supply voltage, V, then the current drawn by R_1, R_2 and R_3 will be V/R_1, V/R_2 and V/R_3 respectively. Therefore the total current is given by

$$I = \frac{V}{R_1} + \frac{V}{R_2} + \frac{V}{R_3} = V\left(\frac{1}{R_1} + \frac{1}{R_2} + \frac{1}{R_3}\right)$$

Therefore the total resistance R is calculated from the formula

$$\frac{1}{R} = \frac{1}{R_1} + \frac{1}{R_2} + \frac{1}{R_3}$$

The Wheatstone Bridge

The Wheatstone bridge network is one of the most important networks we have in electronic circuitry, especially when it is extended to measuring techniques. We shall be referring to this circuit arrangement in several places throughout the book when describing such apparatus as electronic thermometers, burglar alarms, colour sensors, fail-safe circuitry, and so on.

Consider first the circuit diagram of Fig. 9(a), which represents a uniform resistor AB connected to a battery. The potential at A is 10 volts positive with respect to B, which is at zero potential. Half way 'up' the resistor from B the voltage must be +5 volts because there is a steady rise in potential throughout the resistor from B to A. Two thirds of the way up from B the potential must be 6·67 V, and so on. In general the voltage between B and any point C along the resistor will depend upon the resistance between B and C. If the resistance between B and C is R_1 then the voltage between B and C will be R_1 times the current, I, through the resistor. However, by Ohm's law, the current I is 10 volts divided by the resistance between A and B, viz. $(R_1 + R_2)$. So we see that the voltage between B and C is $R_1 \div (R_1 + R_2)$ of the total voltage (10 volts); that is to say, the fraction of the total voltage that appears between BC is the resistance between B and C divided by the total resistance.

If now a second resistor, DE, is connected in parallel with AB, as in Fig. 9(b), a similar argument can be applied to show that the voltage between

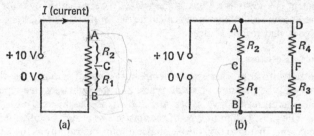

(a)   (b)

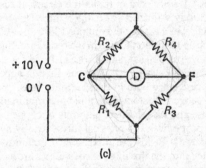

(c)

Fig. 9. The Wheatstone bridge arrangement. At balance the potentials at C and F are the same. The detector, D, then shows a null reading. The balance condition is $R_1/R_2 = R_3/R_4$.

E and F is $R_3/(R_3 + R_4)$. Now a special case arises if the voltage between B and C is the same as that between E and F, because C and F will then both be at the same potential; in other words, the voltage (i.e. potential difference) between C and F will be zero. This occurs when

$$\frac{R_1}{R_1 + R_2} = \frac{R_3}{R_3 + R_4}$$

If we turn both fractions upside down we have

$$\frac{R_1 + R_2}{R_1} = \frac{R_3 + R_4}{R_3}$$

so

$$\frac{R_1}{R_1} + \frac{R_2}{R_1} = \frac{R_3}{R_3} + \frac{R_4}{R_3}$$

But $R_1/R_1 = 1$ and $R_3/R_3 = 1$, so taking 1 from each side

$$\frac{R_2}{R_1} = \frac{R_4}{R_3} \quad \text{or} \quad \frac{R_1}{R_2} = \frac{R_3}{R_4}$$

If we now connect a voltmeter between C and F it will register zero volts. It is customary to draw the circuit of Fig. 9(b) as shown in Fig. 9(c). Such an arrangement is known as a **Wheatstone bridge**. We see that if any three resistances are known accurately we can calculate the fourth. When $R_1/R_2 = R_3/R_4$, the bridge is said to be balanced.

Non-Linear Resistors

These fall into four classes: the temperature-sensitive type, the voltage-sensitive type, the pressure or strain types, and the light-sensitive types.

The temperature-sensitive types are known as **thermistors**, i.e. *therm*al res*istors*. The most common type of thermistor has a high negative temperature coefficient; that is to say, the resistance diminishes rapidly as the temperature increases. (The resistance of wirewound resistors increases as the temperature increases, although not by very much in a well-made component used for electronic purposes.) Within certain limits the electrical resistance of a thermistor is almost entirely a function of temperature. This temperature dependence is so great that over a range of say minus 100°C to +600°C there may be a change of as much as 10 million to one in the resistance. Such high sensitivity to temperature changes as well as small size and rugged construction, make the thermistor an unusually sensitive transducer that is superior to both the thermocouple and resistance bulb in many applications.

Thermistors are semiconductors of ceramic material made by sintering the oxides of manganese and nickel. Small amounts of copper, cobalt or iron are added to vary the properties. The physical forms readily available are beads, discs, washers and rods. Beads are often used as clinical thermometers, in which they are sealed into the tips of glass rods. Discs are a convenient form of thermistor to use when measuring the temperature of flat surfaces.

The available resistance ranges vary according to the type, but with most thermistors it will be found that the resistance falls at about 3 to 5 per cent per degree C. At room temperatures, say 20°C, a rod-type of thermistor may have a resistance of 200 ohms, which will reduce to 5 or 6 ohms when the temperature is increased to 100°C. The small-bead types used in clinical thermometry may have a resistance of 100 kΩ at room temperature, falling to perhaps only 50 kΩ at about 35°C.

Practical Uses

Basically every thermistor application is one of temperature measurement or the detection of temperature changes. In use, the thermistor is operated as an 'externally heated' unit or a 'self-heated' unit. With **externally heated** units changes of ambient or contact temperature produce corresponding changes of current or voltage; these changes are then 'processed' by electronic apparatus. In such an application thermistors are ideal for temperature measurement, control or compensation. **Self-heated** units are so called because they make use of the heating effect of current flowing through them. The heating effect controls the resistance value. Typical applications when used in this mode are voltage regulators, the stabilization of the amplitude of the output voltage from oscillators, microwave power meters, automatic liquid level controls, vacuum gauges and gas analysers.

The simplest type of electronic thermometer using a thermistor is shown in Fig. 10(a). The preset resistor is set to produce a suitable deflection at room temperature. Subsequent changes in temperature alter the resistance of the thermistor and this results in a change of current through the meter. By using known temperatures, the meter can then be calibrated directly in degrees

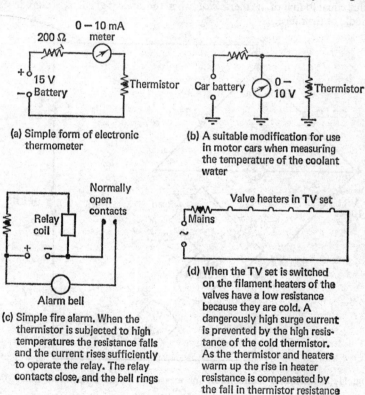

(a) Simple form of electronic thermometer

(b) A suitable modification for use in motor cars when measuring the temperature of the coolant water

(c) Simple fire alarm. When the thermistor is subjected to high temperatures the resistance falls and the current rises sufficiently to operate the relay. The relay contacts close, and the bell rings

(d) When the TV set is switched on the filament heaters of the valves have a low resistance because they are cold. A dangerously high surge current is prevented by the high resistance of the cold thermistor. As the thermistor and heaters warm up the rise in heater resistance is compensated by the fall in thermistor resistance

Fig. 10. Applications of thermistors.

Celsius. An alternative method of measurement is to measure the voltage changes across the thermistor, as in Fig. 10(b). Such an arrangement is very convenient in motor cars since both the meter and the thermistor can be 'earthed' to the chassis. The circuit is suitable for both positive and negative earth systems; care should be taken to connect the meter with the correct polarity. The thermistor can conveniently be a disc type bonded to a suitable point on the radiator with a thin layer of Araldite. Thermistor currents should be limited to a low value to prevent self-heating. The circuits of Fig. 10(c) and 10(d) show two other applications of thermistors.

The simple circuits of Fig. 10 suffer from some disadvantages: they are not very sensitive, cannot give a zero reading and have a non-linear scale (i.e. equal changes in temperature do not bring about equal changes of meter needle deflection). Such disadvantages can be largely overcome by using the thermistor in one of the 'arms' of a Wheatstone bridge. It can be shown that by connecting a resistor in parallel with a thermistor a more linear response can be obtained. The 'padding' resistor, as it is called, should have a resistance value equal to that of the thermistor, when the latter is at a temperature in the middle of the range.

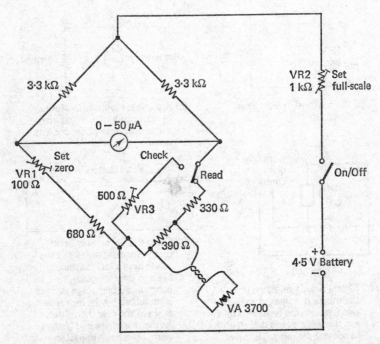

Fig. 11. A thermistor thermometer for measuring in the temperature range 0°C to 100°C.

Fig. 11 shows a suitable circuit for measuring temperatures in the range from that of melting ice (0°C) to that of boiling water (100°C). A miniature bead thermistor is used which has a very small heat capacity; the thermistor thus responds very rapidly to changes of temperature. The instrument should be calibrated by inserting the thermistor into melting ice. VR1 is then adjusted so that the bridge is balanced. This is observed when no current passes through the meter. Adjusting VR1 is equivalent to setting the zero. Then with the thermistor placed just above boiling water, VR2 is adjusted until a full-scale deflection is obtained. VR2 can subsequently be adjusted from time to time to compensate for the change in battery voltage. VR3 is a 'check calibration' control and should be set to produce a full-scale deflection when the switch is in the 'check' position. VR2 can then be adjusted at any future time, as stated above. The scale will be found to be almost linear, thus making the reading of intermediate temperatures easy.

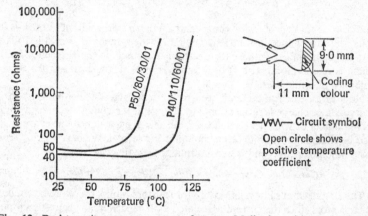

Fig. 12. Resistance/temperature curves for two Mullard positive temperature coefficient thermistors.

Thermistors with positive temperature coefficients are available. Fig. 12 shows how the resistance of such a device varies with temperature. At normal temperatures the resistance is low and presents little interference with the current flow. Above a certain temperature, known as the Curie temperature, the resistance suddenly rises to a large value, thus preventing excess current flow. The main use of positive-temperature-coefficient thermistors is to protect current-carrying circuits from damage due to excessive rises in temperature. For example, an electric motor may be subjected to dangerous rises in temperature in the event of sustained overloading, a locking of the rotor arm or a blocking of ventilation ducts. Such a temperature rise can be prevented by including in the power lead a suitable thermistor. Often these positive-temperature-coefficient thermistors are embedded in the motor windings.

The **voltage-sensitive** type of non-linear resistor is a ceramic-like resistance material with the unusual property of being able to change its resistance in accordance with the applied voltage. The component is formed by dry-pressing silicon carbide with a ceramic binder and firing at about 1,200°C. The usual shape takes the form of discs or rods. The voltage-sensitive resistor is used as a surge-limiter and voltage stabilizer, and also for producing special waveforms.

Capacitors

Capacitors are components that have the ability to store electric charge. A capacitor consists of two conductors in close proximity separated by an insulator called the **dielectric**. If a potential difference, V, is established across the dielectric by connecting the conductors to the terminals of a battery, a charge, q, is stored. Since the charge stored is proportional to the applied voltage, doubling the voltage doubles the charge, trebling the voltage increases the charge three times, and so on. We see that the ratio of the charge to the voltage is constant, i.e.

$$\frac{q}{V} = C \text{ (a constant)}$$

The constant is known as the **capacitance**. When q is in coulombs and V is in volts, the capacitance is in farads. The farad (the symbol for which is F) is a very large unit of capacitance for electronic purposes, so the **microfarad** (μF), which is one millionth (10^{-6}) of a farad, is used. The **picofarad** (pF) is one millionth of a microfarad.

Now in an electronic circuit we are far more interested in current than charge. Since charge is proportional to voltage, if the voltage is changing at a given number of volts per second then, for a fixed capacitance, the charge will also be changing, the units being coulombs per second. We have already seen that coulombs per second is current in amperes; thus we have

Current = *Capacitance* $\times$ *Rate of Change of Voltage*

The simple Ohm's law relationship for resistors cannot therefore be applied to capacitors since the current associated with capacitors is proportional not to the applied voltage, but to the rate of change of voltage. Of course, no current can actually pass through a capacitor (as it can through a resistor) because of the presence of the dielectric, which is an insulator. However, if the voltage applied to a capacitor is varying we can measure varying currents in the wires leading to the capacitor. Let us see how this can be.

Consider the sudden application of a steady voltage to the plates of a capacitor which previously was uncharged (Fig. 13). As soon as the switch is closed electrons from the upper plate, A, are attracted to the positive side of the battery. Since the current in a series circuit is everywhere the same, the number of electrons flowing out of the upper plate equals the number of electrons repelled by the negative terminal on to the lower plate, B. The upper plate therefore becomes positively charged and the lower plate becomes negatively charged. The electrons from the upper plate come from the atoms of the material from which the plate is made. (In a conductor at room temperatures many millions of electrons are in free random motion in the conductor, much as gas molecules are in random motion in a container filled with gas.) Whilst this charging action is taking place electrons are passing along the leads to the battery, so any current detector would reveal a current in these wires even though no current is actually passing through the dielectric.

After a short time, depending upon the size of the capacitor and the resistance of the battery and leads, the voltage between the capacitor plates rises, for all practical purposes, to that of the battery. When this is achieved the flow of electrons stops, i.e. the current falls to zero. It is important to note that the *maximum* voltage across the capacitor occurs at a time when the current falls to *zero*. The maximum current flows at a time when the capacitor is uncharged. Since charge = voltage × capacitance, then for a fixed capacitance when the charge is zero the voltage across the capacitor is zero. The astute reader may well ask what has happened to the voltage being applied by the battery. The answer is that initially, when the charging rate, and hence the

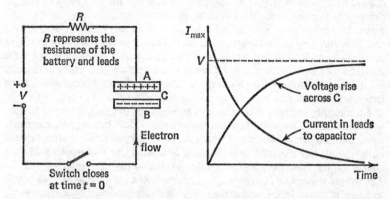

Fig. 13. The charging of a capacitor from a source of steady voltage, V. Note that the current is a maximum when the voltage across C is zero, i.e. when C is uncharged. The current falls to zero when the voltage rises to a maximum. The voltage across the resistor (being the product of the current and the resistance) has the same waveform as the charging current. The sum of the instantaneous voltages across C and R is equal to the supply voltage, V.

current, is greatest, all of the battery voltage is dropped across the resistor R. As the charge on the capacitor builds up so will the voltage across the plates. The battery voltage is then shared between the resistor and the capacitor. Eventually, when the charging is complete, the current falls to zero. There can then be no voltage developed across the resistor, and all of the battery voltage appears across the capacitor.

If the switch (shown in Fig. 13) is now opened, the capacitor remains charged because there is no way for the electrons on plate B to be returned to plate A.

The capacitor thus stores the energy received from the battery and holds it until it is released. (We are assuming, of course, that the dielectric is a perfect insulator. In practice some leakage across the dielectric does occur.) If now a conductor is connected across the plates of the charged capacitor, electrons flow from B to A, thus returning the capacitor to its original uncharged state. The energy stored is dissipated in the form of heat in the conductor since the latter has some resistance.

If the battery is replaced by some source of alternating voltage, then the supply voltage will be constantly changing. If the change is a sinusoidal one, as it is on the alternating current mains supply, then the capacitor will be constantly being charged and discharged—see Fig. 14. There will be a constant ebbing and flowing of electrons in the wires from the supply up to the capacitor, and thus an alternating current will be flowing. When steady voltages are applied (the so-called direct current case), once the initial charging is complete no further current flows. The capacitor thus acts as an open circuit because it offers an infinitely great resistance to the steady applied voltage. This is because the dielectric is an insulator. However, in the alternating current (a.c.) case we obtain an alternating current in the lead wires so long as the supply is connected. No electrons, of course, actually flow through the dielectric of the capacitor; the latter is charging to one polarity, discharging and then charging to the reverse polarity, but so far as the rest of the circuit is concerned an alternating current flows just as it would if the capacitor were replaced by a resistor. There is, however, one extremely important difference. If an actual resistor were used the maximum current would occur at the same time that the applied voltage reaches its maximum value. The supply current and the supply voltage are then said to be **in phase** (meaning in step or in time with each other). We have seen that this does not occur in a capacitor because the maximum current flows at a time when the charging or discharging rate is a maximum; but the maximum charging or discharging rate occurs when the charge on the capacitor, and hence the voltage across the capacitor, is zero. The supply voltage and supply current are therefore out of step—they are said to be **out of phase**. In the case of sinusoidal voltages produced by a dynamo the voltage variations go through a complete cycle during the time that the armature rotates by a complete turn, i.e. travels through an angle of 360°. From Fig. 14 we see that the voltage and current waveforms are out of step by a quarter of a complete cycle; they are said to be 90° out of phase.

The opposition to current flow offered by a resistor in a circuit is measured in terms of the resistance in ohms. The resistance is the same whether the voltage source is steady or alternating. Even if the frequency of the supply voltage varies the resistance remains the same. (At extremely high frequencies this statement is not quite true because the resistor does not act as a pure resistance.) In the case of a capacitor, however, the opposition to flow, which is called the **reactance** of the capacitor, depends not only on the size of the capacitor—the bigger the capacitance the less is the reactance—but also on the frequency. At zero frequency, the d.c. case, the reactance is infinitely great. As the frequency is increased from zero the opposition to flow becomes less and less. These factors are summarized neatly in a formula for reactance, X_C. It can be shown that

$$X_C = 1/(2\pi f C)$$

The 2π arises because we are dealing with sinusoidal waveforms. One complete cycle of the supply voltage is produced when the rotor of the dynamo turns through 360°, i.e. 2π radians. We see therefore that the reactance is inversely proportional to the frequency and inversely proportional to the capacitance.

Three factors affect the capacitance of a capacitor: the area of the plates, the distance between the plates and the nature of the dielectric material. For a given dielectric the capacitance is proportional to the area of the plates and inversely proportional to the distance between them. Associated with the insulating material used between the plates is a constant known as the **dielectric constant**. For air (strictly a vacuum) the dielectric constant is 1 whilst for mica and paraffin-waxed paper it is 6 and 2·2 respectively. The greater the dielectric constant the greater the capacitance will be.

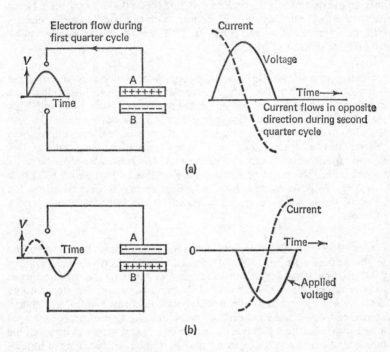

(a)

(b)

Fig. 14. Application of a sinusoidal voltage to a capacitor. (a) After the regular ebbing and flowing has been established (i.e. the so-called steady-state condition) the voltage and current waveforms are as shown on the right. When the voltage is zero the current is a maximum. At maximum voltage the current is zero and the capacitor is fully charged. As the applied voltage falls, the voltage on the capacitor causes current to be returned to the source. (b) Current and voltage waveforms during the next half-cycle of applied voltage.

The choice of the dielectric is not made by considering the dielectric constant alone. The **dielectric strength** (i.e. the voltage per given thickness before electrical breakdown occurs) must be considered, too. Economic and other factors are also involved; mica, for example, is an excellent dielectric, but the cost of making a capacitor of several hundred microfarads would be prohibitive.

No dielectric is perfect, and some leakage is inevitable. For mica and some plastic foils the leakage is extremely small; but in electrolytic capacitors the leakage may be several milliamps. In some circuit positions, e.g. in power supplies, the leakage is not important, but in other locations leakage may be of paramount importance.

The physical appearance and properties of a capacitor vary a good deal depending upon the nature of the 'plates' and the dielectric material. Values of capacitance from 1pF to several thousands of microfarads are readily available. One classification of capacitance depends upon the dielectric used. Hence we have general-purpose paper capacitors, mica capacitors, ceramic, electrolytic, polystyrene and polycarbonate types. The main physical constructional features are shown in Fig. 15.

General-purpose capacitors use paper impregnated with wax (or oil) as the dielectric (Fig. 15(b)). Two long rectangular aluminium foils separated by a slightly wider strip of impregnated paper are rolled up like a Swiss roll. They may then be inserted into an insulating tube and sealed at the ends. Connecting wires to each foil are brought out separately from each end; alternatively the rolls may be encapsulated in some form of plastic insulating material. This type of capacitor is relatively inexpensive and has a reasonable capacitance-to-volume ratio. The range of capacitance available is from about 100 pF to 1 or 2 μF. Working voltages vary, but values up to 600 V are common. The value of the capacitance is clearly stated on the body together with the tolerance.

Capacitors using mica dielectrics have a capacitance range from a few pF up to about 50 nF—1 nF (nanofarad) is equal to 1000 pF, which is the same as 0·001 μF. Precision mica capacitors are used as standards and can be adjusted to have values within 0·01 per cent of the nominal values. The higher values are obtained by stacking several metal foils interleaved with mica sheets. Tolerances of 5, 2 and 1 per cent are common. This type of capacitor is protected by sandwiching between insulating boards and waxed to keep out the moisture; alternatively the capacitor may be housed in a moulding of insulating material.

Ceramic dielectric capacitors are made with three different values of dielectric constant, k. The low-k types are made from steatite and have excellent high-frequency performance. The medium-k types have a negative temperature coefficient and are used as temperature-compensating capacitors in tuned circuits. The high-k types with dielectric constants up to 1,200 are used where space is at a premium. Unfortunately, the latter type is very temperature-sensitive. The capacitance value of ceramic capacitors is usually indicated by a 'band+4 dot' system, a 'five-dot' system or marking in figures. No standard system is used throughout European countries and America, and some confusion can arise in reading the markings. In general, the first and second dots represent the significant figures, the third the decimal multiplier and the fourth the tolerance. The band is used to indicate the temperature coefficient. The best way out of the difficulty is to buy capacitors from a reliable source using a known code.

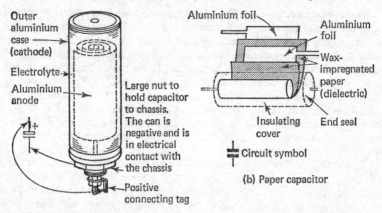

(a) The construction of a 'wet type' of electrolytic capacitor

(b) Paper capacitor

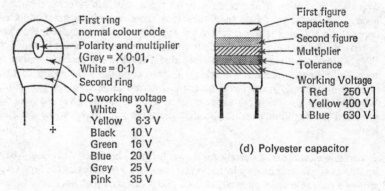

(c) Tantalum capacitor: with dot facing viewer, positive lead is to the right

(d) Polyester capacitor

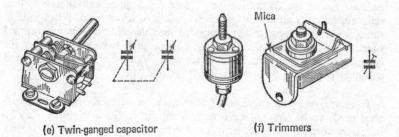

(e) Twin-ganged capacitor

(f) Trimmers

Fig. 15. Various forms of capacitors.

Electrolytic capacitors (see Fig. 15(a)) are used when large values of capacitance are required in a physically small component. Typical applications are in power supply smoothing filters, by-pass capacitors and in what are called biasing and coupling circuits. (These terms are explained in the appropriate chapters on power supplies and amplifiers.) In the low-voltage range capacitances up to thousands of microfarads are readily available. The tolerances are fairly wide, usually being -20 to $+50$ per cent. The ordinary plain foil types (the so-called 'wet' electrolytics) have a thin rectangular sheet of aluminium wound in a spiral so as to be conveniently housed in a metal cylinder. The spiral sheet, previously etched with acid to increase its surface area, is immersed in an electrolyte, which is usually a solution of ammonium borate. The solution, like all electrolytes, can easily conduct electric current. During a 'forming' process a thin film of aluminium oxide is deposited on the spiral, which serves as the dielectric. The oxide layer is electrically very strong, i.e. it can withstand quite high voltages. An incidental advantage with this type of dielectric is that it is self-healing. If an electrical breakdown occurs because of the application of not too great a voltage overload then, on removing the overload, the action of the electrolyte on the aluminium reforms the oxide layer. Care must be taken to ensure that these capacitors are connected correctly in the circuit since, unlike the types previously discussed, they are polarized. The manufacturer clearly marks which connection must be made to the positive potential. The capacitance value is marked on the body in figures together with the maximum working voltage conditions. Needless to say, alternating voltages must never be applied across an electrolytic capacitor. Variable voltages which are superimposed upon a steady level are allowed provided the variations do not produce negative potentials on the positive lead.

Variable capacitors generally take two forms. The first is probably best known as the tuning agency in radio receivers (see Fig. 15(e)). This type consists of a movable set of specially shaped plates or vanes that interleave with a set of fixed plates. The movable set of plates is rotated by means of a spindle. The two sets are electrically isolated by the dielectric, which very frequently is air. In some of the older types of tuning capacitors thin sheets of flexible insulant act as the dielectric; the volume for a given capacitance is thus reduced, which is an advantage when space is at a premium. When used in radio sets, two or more of the air-spaced variable capacitors may be ganged together so that the rotation of only a single shaft is required to alter the capacitance in several circuits simultaneously. Values for a single air-spaced capacitor commonly lie between 50 pF and 500 pF.

Trimmers, or preset capacitors (see Fig. 15(f)), are used when the capacitance needs to be changed only very infrequently. These trimmers may be miniature versions of the larger types of variable capacitor; alternatively they may be of the compression type where, by turning a screw, the metal foils and dielectric sheets are compressed to a greater or lesser degree. The resulting change in distance between the metal foils alters the capacitance.

Capacitors in Parallel and Series

When capacitors are connected in parallel the same voltage exists across each capacitor. Each capacitor therefore stores a charge proportional to the capacitance. The total charge is the sum of all the charges. The parallel arrangement therefore has a capacitance which is equal to the sum of the individual capacitances, or

$$C = C_1 + C_2 + C_3, \text{ etc.}$$

Connecting capacitors in series results in a reduction of the total capacitance to a value less than that of the smallest capacitor in the chain. The formula for the total capacitance is given by

$$C = \cfrac{1}{\cfrac{1}{C_1} + \cfrac{1}{C_2} + \cfrac{1}{C_3} \cdots}$$

Inductors

Inductance is that property of an electric circuit which tends to prevent any change of current in that circuit. Devices having the primary function of introducing inductance into a circuit are called **inductors**. Inductance can be considered as the electrical equivalent of inertia. For example, a flywheel on a motorcar or steamroller tends to prevent any change in speed. The flywheel is deliberately made heavy so that when it rotates it stores considerable energy. When the road wheels meet an incline or bump, the speed of the vehicle is maintained because it is difficult to change the speed of rotation of the fly-wheel. Usually the engine cannot deliver all the necessary energy in a short space of time to overcome the obstacle; the necessary energy to maintain the speed is taken from the flywheel.

All inductors (which are the electrical analogues or equivalents of flywheels) consist of coils of insulated wire wound on to suitable bobbins or formers. Although the core of the coil may be air, it is more usual to concentrate the magnetic flux in the core by using suitable ferromagnetic substances; magnet-ically soft iron (i.e. iron which is easy to magnetize or demagnetize) and ferrites are often used. Since inductors oppose changes of current in a circuit, one of their functions is to present a large opposition to the flow of alternating current whilst simultaneously presenting very little opposition to the flow of direct current. The small opposition to the flow of steady or direct current comes from the resistance of the coil, not its inductance. We find this property to discriminate against the alternating current very useful in power supplies. Such supplies are used to convert the alternating voltages of the mains electri-city supply to a steady or direct voltage suitable for energizing electronic equipment. We will need to refer to this again when we come to see how power supplies are designed.

To understand the way in which inductors present an opposition to alter-nating currents, but not to direct currents, we need to recall some basic facts about electromagnetism. An electric current flowing through a conductor has a magnetic field associated with it. This we can demonstrate by passing the conductor through a horizontal piece of card on which iron filings are sprinkled. With no current, a tapping of the card merely produces a random orientation of the filings. When we pass a direct current through the conductor and tap the card again, the iron filings take up a pattern of rings concentric with the conductor. This shows the shape of the magnetic field. It is usual to visualize the magnetic field by using the concept of magnetic flux lines, i.e. paths that would be taken by a fictitious isolated north pole if it were free to move. Such flux lines are shown in Fig. 16.

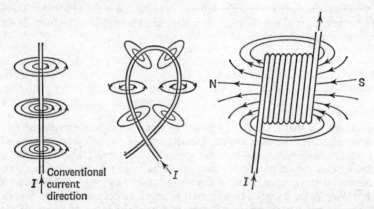

Fig. 16. Lines of force representing magnetic fields around current-carrying wires in which the current is steady. Steady magnetic fields associated with conductors do not, however, induce currents in the wires. To induce a current the field must be varying.

Whilst steady currents produce magnetic fields, Faraday showed that the converse is not true; to induce a current in a conductor, it is necessary for the magnetic field to vary. Consider now what happens when we bend the current carrying wire into a loop (Fig. 16). We see that there is a concentration of the flux inside the loop. If we now wind the wire into the form of a coil the flux takes on the shape shown in the diagram. If the current is cut off the magnetic field disappears. By placing a magnet in the coil, experiment shows that no voltage is produced between the ends of the coil provided both the coil and magnet are stationary. If, however, the magnet is moved relative to the coil, a voltage is produced across the ends of the coil. The voltage is found to be proportional to the rate at which the flux is changing in the coil. This means that if the coil is connected to an external circuit a changing current is produced in the circuit and the coil. This current itself gives rise to a magnetic field; this field opposes the magnetic field inducing the current.

Instead of moving a magnet into the coil let us now apply a voltage to the ends of the coil. If the voltage is steady the only opposition to the current is the resistance of the wire forming the coil. If the voltage is varying, however, a continual change occurs in the current in the coil, which in turn produces a changing magnetic field. As the magnetic field grows, an increasing flux threads the coil. The rate of increase depends upon the frequency of the applied alternating voltage. When a magnetic field grows inside a coil an increasing magnetic flux threads the coil. Such a changing magnetic field induces a voltage that opposes the applied voltage. The induced voltage is called a **back electromotive force** (usually abbreviated to **back e.m.f.**). There is thus some opposition to the establishment of the magnetic field, and hence an opposition to the rise of current in the coil.

So far we have considered what is happening when the varying applied voltage is increasing. When the applied voltage diminishes the magnetic field collapses. This collapsing field induces a voltage which tends to keep the current at its maximum value. Whether the applied voltage is rising or falling therefore the inductor opposes changes of current in the circuit. This opposition is proportional to the size of the inductor and the frequency of the applied voltage. The opposition is called the **inductive reactance** and is often represented by the symbol X_L. It can be shown that $X_L = 2\pi fL$, where f is the frequency in hertz (cycles per second) of the supply voltage and L is the inductance in henries. When, in an inductor, a current that is changing at the rate of 1 ampere per second produces a back e.m.f. of 1 volt then the inductor has an inductance of 1 henry.

Fig. 17 shows the constructional features of two common types of inductor.

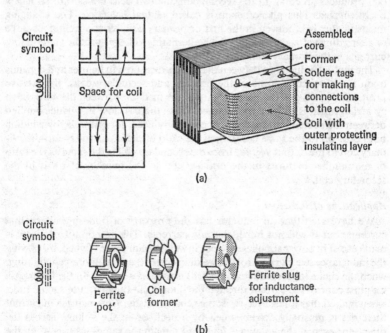

(a)

(b)

Fig. 17. Two forms of inductor. (a) Construction of an inductor for power frequencies. The shape of the laminations is shown on the left. Either combination may be used. (b) Essential parts of a 'pot-core' inductor.

For use at low frequencies (especially 50 Hz and 100 Hz) iron cores are used to increase the inductance of a given coil. The iron cores are laminated, each lamination being coated on one side with an insulating material in order to prevent eddy currents in the core. Such eddy currents can cause severe losses of energy. At radio frequencies even the use of insulated laminations does not prevent severe losses. The cores of high-frequency inductors are therefore made from ferrites or iron dust held in a suitable insulating material. Ferrites are special oxides of magnesium, manganese, zinc or nickel; they are non-conducting so far as electric currents are concerned so that eddy currents at high frequencies are largely avoided. Permeabilities of several hundred up to about 1,200 are readily obtained. **Permeability** is a measure of the increase in inductance that results from the use of a core. For example, if an air-cored coil has an inductance of 1 mH (millihenry, i.e. one thousandth of a henry), then by using a ferrite core of the type shown in Fig. 17(b) with a permeability of say 400, the inductance will be increased to nearly 400 mH. We have to say 'nearly' because some losses are bound to occur.

Mutual Inductance

If two coils are placed close together so that a varying magnetic field in one coil produces an e.m.f. in the second coil, the two coils are said to be **inductively coupled.** This phenomenon is called **mutual induction**. The changing magnetic flux is produced in the first or primary coil, by connecting a source of alternating voltage to the coil. Such alternating voltages produce alternating currents in the primary coil.

The mutual inductance between the coils is measured in henries and depends upon the number of turns in the primary and secondary coils, the relative positions of coils and the permeability of the medium between the coils. Such an arrangement is called a **transformer**. When there is close magnetic coupling between the primary and secondary coils (produced, for example, by winding both coils on to the same core), then the ratio of the input voltage to the primary coil to the output voltage from the secondary coil is the same as the ratio of the number of turns in the primary coil to the number of turns in the secondary coil.

Applications of Inductors

We have seen that an inductor has the property of opposing alternating currents, but it will not impede steady currents. This important property is made use of in power supplies, in electric welding equipment (when, by varying the inductance, we can control the welding current) and in fluorescent lighting, when the **choke** (as an inductor is called) is used as a ballast. In the latter case we must control the current through the fluorescent tube because such a tube, when ignited, has a negative-resistance characteristic. Any increase in current therefore is normally accompanied by a decrease in the voltage across the tube. However, if the voltage is forced to remain the same—as it is when the tube is being driven by the mains—a destructive rise in current would result. Such a rise can be prevented by using a choke in series with the tube. The inductor performs an additional function in fluorescent tube apparatus. The properties of the fluorescent tube are such that in order to ignite the tube a higher voltage is required than that which will maintain the gas discharge. Since the tubes are designed to run at mains voltages some mechanism must produce a high 'striking' voltage. The most efficient way of doing this is to have a starter which first allows the current to be established in the choke and then suddenly breaks the circuit. When the current is quickly cut off in this way the flux in the choke collapses rapidly. A very high voltage is therefore produced across the choke as a consequence of the rapidly changing flux. This high voltage is sufficient to ignite the tube which thereafter 'burns' at the lower maintaining voltage.

The Response of Circuits Containing Passive Components

Sine Waves

The generation of voltages and currents with a sinusoidal waveform is usually carried out in a laboratory by a piece of electronic equipment called an **oscillator**. Later in the book a whole chapter is devoted to the subject of oscillators, but here we shall content ourselves with the dynamo type of sine wave generator shown in Fig. 18. The alternating current supply mains uses a

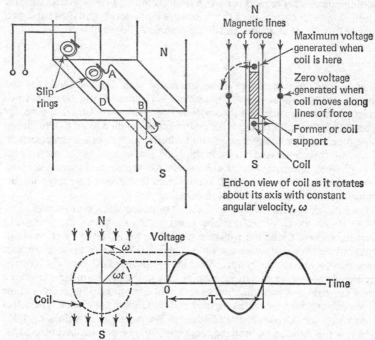

Fig. 18. The generation of sine waves in a simple dynamo: 1 revolution = rotation of 2π radians $\therefore \omega T = 2\pi$

$$T = \frac{1}{f} = \frac{2\pi}{\omega} \therefore \omega = 2\pi f$$

sinusoidal waveform because this type of variation is easy to generate and avoids design difficulties in the distribution equipment used for high powers.

The simple sine wave generator, shown diagrammatically in Fig. 18, consists of a rectangular turn of wire which is rotated in a magnetic field. When the wires along the length of the coil, AB and CD, are travelling parallel to the direction of the lines of force then the induced voltage is zero because there is no cutting of the lines of magnetic flux. When AB and CD are moving at right angles to the lines of force there is a maximum rate of cutting of the lines of force and hence the voltage induced in the coil is a maximum. As the coil continues to rotate, the induced voltage falls to zero, and thereafter will increase in magnitude, but be of opposite polarity. Provided that the coil is rotated at a fixed rate, the voltage between the slip-rings has a sine waveform. The fixed rate of rotation (in radians per second) is usually given the symbol ω. One revolution, which is equivalent to a rotation of 2π radians (360°), produces one complete cycle of the waveform. The time taken to complete one cycle is known as the **period** or **periodic time**, T. If the frequency of the waveform is f, this means that f rotations of the coil take place in 1 second. Since each rotation is equivalent to an angle of 2π radians then the total angular rotation must be $2\pi f$ radians per second; this is the angular rate of rotation ω; therefore $\omega = 2\pi f$. Also, since f rotations are made per second then 1 rotation is made in $1/f$th of a second; this is the period, T. So we have

$$T = \frac{1}{f} \quad \text{and} \quad \omega = 2\pi f$$

One of the most important terms associated with sine wave voltages and currents is the **root-mean-square (r.m.s.) value**. The need to use the r.m.s. value arises when we consider powers associated with this waveform. Suppose, for example, that we wish to know the power being supplied to an electric fire or some other heating equipment, or to an electric motor. In the direct current (d.c.) case the power is given by multiplying the voltage by the current, i.e. $P = VI$. With alternating current, however, the voltage and current are constantly varying, so we must find some figure that effectively represents the sinusoidal voltage for power calculations.

Since power is proportional to the square of the voltage or current ($P = I^2R = V^2/R$), we therefore find the mean of the square of the alternating quantity and then take the square root; hence we have the r.m.s. value. It turns out that for sine waves the r.m.s. value is 0·707 times the amplitude or peak value. Although this result is obtained by advanced mathematics it can be confirmed by experiment. In such an experiment it can be discovered that a direct current gives two times the heat output of an alternating current of the same peak value. In order that we may perform simple Ohm's law type calculations with a.c. circuits it is necessary to divide the peak values by 1·414 (which is the same as multiplying them by 0·707). We thus obtain the r.m.s. value (1·414=$\sqrt{2}$).

So far as the heating effect is concerned an alternating current of 1 amp (r.m.s.) is equivalent to a direct current of 1 amp. This is why most a.c. measuring instruments are calibrated by the manufacturers in terms of r.m.s.

values. Unless otherwise stated, a alternating voltage of, say, 250 volts implies that 250 volts r.m.s. is meant.

Response of *C*, *R* and *L* Circuits to Signals

We have already discussed how, when we apply a voltage with a sine waveform to a resistor, the voltage and current reach their maximum values at the same time. With a capacitor, however, once the sinusoidal voltage and current waveforms have been established, the current reaches its maximum value a quarter of a period before the voltage reaches its maximum value. For inductors the current reaches its maximum value one quarter of a period later than the time when the voltage attains its maximum value. These results are summarized in Fig. 19, together with the expressions or formulae for the opposition to current flow imposed by the three components.

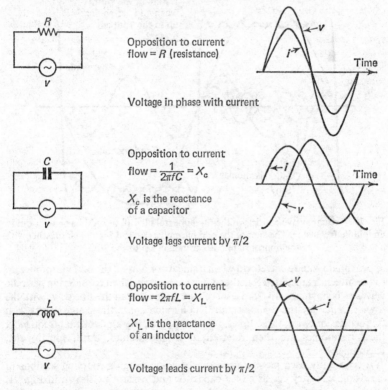

Opposition to current flow = R (resistance)

Voltage in phase with current

Opposition to current flow = $\dfrac{1}{2\pi fC} = X_c$

X_c is the reactance of a capacitor

Voltage lags current by $\pi/2$

Opposition to current flow = $2\pi fL = X_L$

X_L is the reactance of an inductor

Voltage leads current by $\pi/2$

Fig. 19. Phase diagrams for the three basic passive components.

Fourier (1768–1830), a French mathematician, was one of the first to realize that all periodic waves could be synthesized (i.e. put together or built up) by combining sine waves of the appropriate amplitude, frequency and phase. Since many of the signals we wish to process with electronic equipment are sinusoidal or periodic, we see that a study of the response of circuits containing combinations of resistance, capacitance and inductance to sinusoidal or periodic waves is very important to those who design or build electronic apparatus. For example, if an amplifier does not perform well when a square-wave signal is applied to the input terminals then the performance of that amplifier is unlikely to be satisfactory when other periodic signals are applied. With experience a skilful experimenter, on applying square-wave signals to the input terminals, can make deductions about the response of an amplifier at high and low frequencies by observing the output waveforms from the amplifier with the aid of a cathode-ray oscilloscope. This is not surprising when it is realized that a square wave contains the fundamental and all the odd harmonics up to an infinitely great frequency. Fig. 20 shows the resultant

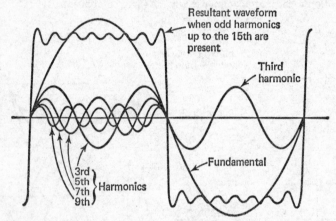

Fig. 20. A square wave consists of the fundamental plus all the odd harmonics out to an infinite frequency. The sum of the odd harmonics up to 15 is shown together with the appropriate odd harmonics up to the 9th.

approximate square waveform when a sine wave plus all the odd harmonics up to the fifteenth are added together. (It is customary, when considering periodic waves as being made up from a set of sine waves, to call the sine wave with the lowest frequency the **fundamental**. The **harmonics** are those sine waves with frequencies of twice, three times, four times, etc., the fundamental frequency; these harmonics are then referred to as the second, third, fourth, etc., harmonics.)

A frequently used passive network for connecting or coupling sections of electronic circuitry consists of a capacitor and resistor as shown in Fig. 21. We must now consider how much distortion of the waveform occurs when various choices of the values C and R are made. A good deal of information can be obtained by observing the effect that various combinations of C and R have on the waveform of a square wave. We need not concern ourselves here just how the square waves are produced; in practice a special type of oscillator is used, but we shall show the production of a square wave from a simple battery/switch arrangement as in Fig. 21. Here the switch is made to change rapidly back and forward between A and B.

If we apply a square-wave voltage to the input of the network, the voltage across the resistor will follow at once the leading edge of the wave. This is because insufficient time has been allowed in the first few microseconds for the capacitor to change its charge. It is essential for the reader to remind himself that the voltage across a capacitor is found by dividing the charge by the

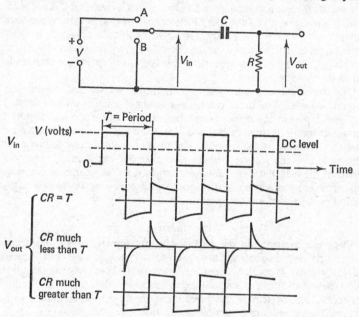

Fig. 21. Waveforms associated with CR circuits of different time constants. Note that the d.c. level is lost in each case.

capacitance. For a given capacitor, if the charge does not change, the voltage across the capacitor cannot change. So if the voltage on the input terminal of the capacitor suddenly jumps from 0 to V volts then the output terminal must also jump to V volts. What happens thereafter depends upon the values of C and R, and the time the input voltage stays at its positive value. During the time that the input voltage is maintained at a steady value, the voltage across the resistor will drop as the capacitor charges via the resistor. (See Fig. 13 and the associated text.)

The rate at which the voltage across R drops depends upon the values of C and R; the greater these values the slower will be the fall of V_{out}. It is not the individual values of C and R that count, but their combination; it is the product CR that is important. CR is known as the **time constant** of the circuit. When C is in farads and R in ohms the time constant is in seconds. When the product CR is about the periodic time of the input waveform we obtain the output waveform that is shown in Fig. 21 next to the statement $CR = T$, i.e. the product CR is approximately equal to the period T. Here the resistance value may be too small, thus allowing a fairly rapid discharge of C, or C may be too small and unable to supply sufficient charge to maintain the voltage across R for a long enough time. As we see, distortion of the input waveform occurs. This distortion becomes particularly severe when $CR \ll T$, i.e. when the time constant, CR, is much less than T. In order to produce the minimum distortion we must make the time constant much larger than T (i.e. $CR \gg T$). Here it should be noted that very little change of charge takes place because either R is very large or C is large (and hence holds a big charge), or both are large. In practice it is usual to make CR at least five, and preferably ten, times the period of the input waveform. The output waveform is then almost identical to the input waveform.

One important difference between the input voltage and output voltage should be noted. The input voltage varies about a mean positive level. This d.c. level is lost in the output waveform because the capacitor transmits only the variations of the input signal and blocks out any d.c. component. We shall find this property of the coupling circuit very useful in our future study of electronic circuits. Readers should note that the waveform is preserved by virtue of the fact that the coupling capacitor does *not* charge or discharge to any significant effect when the time constant is long compared with the period of the waveform being transmitted.

Summary

Electronic systems are the means whereby signals from transducers are suitably processed to activate output devices. The signals are usually periodic, i.e. they have a regular, repetitive waveform. Electronic apparatus is a combination of resistors, capacitors, inductors, diodes, transistors and other semiconductor devices. These basic components can be combined to form many different types of electronic apparatus.

Linear resistors obey Ohm's Law. A resistor has a resistance of 1 ohm if, when a voltage of 1 volt is applied across it, a current of 1 ampere flows through it. Resistances in series are merely added so that $R_{\text{total}} = R_1 + R_2 + R_3$, etc. Resistors in parallel are found by $1/(R_{\text{total}}) = 1/R_1 + 1/R_2 + 1/R_3$, etc.; thus if two resistors are connected in parallel their effective resistance is the product of the two resistors divided by the sum.

A capacitor has the property of storing charge. When 1 coulomb is stored by a capacitance of 1 farad then the voltage across the capacitor is 1 volt. Capacitances in parallel are added, thus $C_{\text{total}} = C_1 + C_2 + C_3$, etc.; capacitors in series have a combined capacitance given by $1/(C_{\text{total}}) = 1/C_1 + 1/C_2 + 1/C_3$, etc. When an alternating voltage is applied to a capacitor an alternating current can be measured in the leads to the capacitor. The alternating voltage divided by the alternating current gives, by the a.c. equivalent of Ohm's Law, a measure of the opposition to the flow of current known as the capacitative reactance, $X_c = 1/(2\pi fC)$. The impedance of a resistor in series with a capacitor is given by $z = \sqrt{[R^2 + 1/(2\pi fC)^2]}$.

Inductance is that property of a circuit that attempts to prevent changes of current in that circuit. A coil has an inductance of 1 henry if the current changing through it at 1 ampere per second induces a voltage across it of 1 volt. **Inductances in series** are added provided there is no mutual inductance between the coils, i.e. $L_{total} = L_1 + L_2 + L_3$. **Inductances in parallel** obey the reciprocal formula $1/L_{total} = 1/L_1 + 1/L_2 + 1/L_3$.

Non-linear resistors follow a more complicated pattern of behaviour. The resistance is not proportional to voltage, but depends strongly on the temperature of the component. The resistance of the majority of thermistors falls exponentially with temperature. With positive temperature coefficient thermistors the resistance rises with temperature.

Sine waves are the simplest waves possible. All periodic waves can be synthesized by combining sine waves of different amplitude, frequency and phase.

Questions

1. Why are resistors manufactured with seemingly odd resistance values such as 47 K, 68 ohms, 2·2 M?

2. A conductor is moved at right angles to a magnetic field and a galvanometer is connected across the ends of the conductor. What will be the effect on the voltage induced if (a) the conductor movement is reversed, (b) if the conductor is moved to and fro within the field and (c) if the conductor is held stationary within the field?

3. Why are the iron cores of power inductors (a) laminated and (b) made of magnetically soft iron?

4. A capacitor of 50 μF is connected across a 10 V d.c. supply. What charge is stored by the capacitor?

5. What factors influence the choice of a given type of capacitor for use in an electronic circuit?

6. List the advantages and disadvantages of electrolytic capacitors as compared with other types.

7. Why is magnetically soft iron not used for the cores of inductors that are operated at radio frequencies?

8. What is meant by the term 'time-constant' in relation to an *RC* circuit?

9. A current is passed through a circuit consisting of a resistor in series with a capacitor. The voltages across the resistor and across the capacitor are measured using an a.c. voltmeter. Why is the voltage across the series combination different from the sum of the individual voltages?

Suggested Further Reading

Basic Electronic Circuits, by a special team of Electronics Training Investigation Team of REME, The Technical Press.

Dummer, G. W. A., *Modern Electronic Components*, Pitman, 1966.

3

SEMICONDUCTOR DEVICES

Resistors, capacitors and inductors, being passive components, are unable by themselves to amplify or make larger the very small signals that are developed by the kinds of transducers associated with electronic apparatus. The need for amplification is so important that a whole chapter later in the book is devoted to the subject of amplification and the associated circuitry. In such circuits the usual amplifying agency today is the **transistor**.

In 1948 the American physicists, J. Bardeen and W. H. Brattain, announced the invention of the transistor, a new type of amplifying device made from semiconducting crystals. Very few at that time could have foreseen the revolutionary developments that were to follow, developments so important and far-reaching as to change the whole outlook of the science and technology of electronics. The physical principles involved in transistor action had been worked out in conjunction with their colleague, W. Shockley; in recognition of their work the three physicists were awarded jointly the 1956 Nobel Prize for Physics.

From the early days of the Second World War until 1950, a very great increase was made in our knowledge of the properties of semiconductors. Semiconductors had, of course, been known for a long time; Faraday in 1833 performed experiments with galena and carborundum; the rectifying properties of certain solids were discovered some two years later. A silicon rectifying diode was known as early as 1906 with the invention of a detector by Pickard, and most readers will be aware of the 'cat's whisker' detectors of early radio sets that made use of the rectifying properties of a metal-to-semiconducting crystal contact. The crystal in this case was usually impure galena, and many a tricky domestic situation arose in the selection and maintaining of the best location of the 'cat's whisker' on the crystal. The invention and development of thermionic diodes and triodes overshadowed the crystal diode, rendering it temporarily obsolete, but prior to and during the Second World War a great deal of work was done on germanium for telecommunication purposes. It is now certain that transistors and allied semiconductor devices will replace thermionic tubes in almost every type of electronic equipment.

The term 'transistor' arises from a combination of the italicized portions of the words *trans*former and res*istor*, since the device is made from resistor material and transformer action is involved in the operation of a transistor. The first transistors were of the **point contact** variety, but two or three years later the **junction** transistor made its appearance. In the early days the point-contact transistor performed better at higher frequencies than the junction types; but improved manufacturing techniques have enabled the junction transistor to establish its superiority in this respect. The vulnerability of point-contact transistors to mechanical shock has precluded the use of this device for all practical purposes; the junction transistor is now so firmly established that we will consider this type alone.

Transistors have several advantages over their thermionic counterparts. They are small and light, being usually less than one gram in weight and less than 1 cm^3 in volume. Most of the bulk is made up of the container and silicone grease or other filler. The actual transistor itself is very small; in micromodule integrated circuits several hundreds of transistors can be fabricated on a single silicon chip about 0·1 inch square, although the single transistor in normal use is somewhat larger than the micromodule type. Transistors can withstand mechanical vibrations and shocks that would ruin a thermionic valve. This is because a transistor is a single-crystal device and therefore has no delicate electrode structure to damage. In addition, there is no vacuum to preserve within a vulnerable glass envelope. Heaters are not required in transistors and thus there is a substantial saving of power; a small amplifying thermionic valve usually requires about 2 watts of heater power. Since no heaters are required, transistors operate immediately the supply is connected, no warming up period being necessary. The supply voltages required are low and rarely exceed 40 V. Voltages from 6 to 24 V are common. As far as can be judged, the life of a transistor that is properly installed and operated within its ratings should be indefinitely long. It is therefore an ideal device for use in remote locations, such as space satellites and submarine telecommunication cables, that need booster or repeater amplifiers along the line.

Problems associated with limited power output, noise, frequency response and operating temperatures are all yielding to intensive research. Nowadays there are very few applications in which a thermionic device is clearly superior to a transistor.

Conductors, Insulators and Semiconductors

The devices discussed in this chapter depend for their action on the controlled flow of electric charges through a suitably prepared crystal. To understand the rectifying and amplifying properties of crystal devices a certain amount of elementary solid-state physics needs to be known.

Our present concept of the nature of matter is that all materials are made of atoms. These atoms are basically electrical and have a small, massive, positive nucleus (consisting of protons and neutrons) which is surrounded by a cloud of electrons. In a neutral atom the number of electrons equals the number of protons. The charge on an electron is equal to that on the proton, but of opposite sign, the electron being negative and the proton positive. After the discovery of the electron by J. J. Thomson in 1897, later attempts by scientists to describe atomic structure led to the idea that the electrons could occupy only specific shells. The planetary system around the sun is often used as an analogy, although readers should note that great theoretical difficulties arise if we try to relate the analogy too closely to sub-atomic particles. Under certain circumstances atoms can be made either to shed or to capture electrons; alternatively, they may share electrons with neighbouring atoms, as is the case in many crystal structures such as diamond, germanium and silicon. The electrons involved are those in the outermost shell, these electrons being responsible for the chemical and principal physical properties of the material. The properties with which we are most concerned in this chapter are those associated with **conduction**. Broadly speaking, materials can be classified as conductors, insulators and semiconductors.

Insulators and Conductors

Some materials, like quartz and mica, will not conduct electricity, whereas copper, silver, aluminium and mercury, as well as many salt solutions, conduct electricity quite easily. Insulators are made from materials whose atoms do not shed electrons, but hold these electrons in a tight bond to the parent nuclei. Since there are no free electrons in the material there are no free charge carriers available to form an electric current, even though the material may have a voltage applied to it. Such materials may only conduct under conditions of extremely high temperatures or voltages, when in general the material disintegrates. Where disintegration does not occur it is found that increases in temperature reduce the resistance. This is because, although practically no free electrons exist at room temperatures, elevated temperatures give sufficient energy to some electrons to enable the latter to become detached from the parent atoms. Glass rods, for example, in a dry, clean state are quite good insulators at room temperatures, but when red hot these rods are found to have an appreciable conductivity.

Solid metallic conductors consist of **poly-crystalline** material. Each tiny crystal in the mass consists of atoms arranged in a regular pattern, called a **lattice**. The outer electrons of atoms of copper or silver, for example, are easily detached, and at ordinary room temperatures many millions of these electrons migrate freely with random motion through the crystals. The main lattice is in constant vibration, however, and consequently there is a high probability that electrons will collide with the vibrating atoms; they also have to cross the barriers at the junctions between the crystallites making up the bulk material of the conductor. It is these collisions that account for the resistance of material. We can see now why the resistance of metallic conductors rises with temperature. Although liquids such as mercury or molten silver, for example, do not possess a crystal structure, the conduction mechanism is very similar to that of solid copper or aluminium.

Conduction in chemical solutions is somewhat different. When many salts or acids are dissolved in water they become ionized. For example, common salt, which chemically is called sodium chloride, consists of crystals containing sodium and chlorine atoms. When dissolved in water these atoms are free to migrate; the sodium atom loses an electron—thus becoming a positive ion—whilst the chlorine atom acquires this electron to become a negatively-charged chlorine ion. The ionization takes place as a result of the salt going into solution and would take place whether or not any external voltages are applied to the solution. If two platinum electrodes are then immersed in the solution and one of them is connected to the positive terminal of a battery whilst the other is connected to the negative terminal, then a migration of the ions takes place (see Fig. 22). The sodium ions are attracted to the negative electrode (**cathode**) and the chlorine atoms are attracted to the positive electrode (**the anode**). On reaching the positive electrode the chlorine atom gives up its electron, which is now free to migrate along the electrode and wire connection to the battery. The chlorine ion now becomes a neutral chlorine atom, which combines with

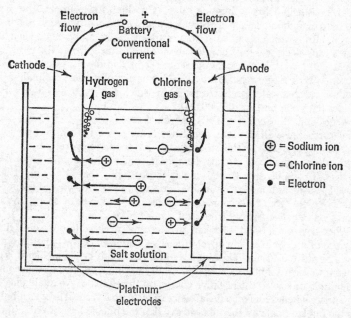

Fig. 22. Conduction of electricity in liquids.

another chlorine atom in the vicinity (by the sharing of electrons in the outer shell) and chlorine gas is produced. At the cathode sodium atoms are produced, but these react with the water present to form sodium hydroxide and hydrogen gas. The production of sodium comes about because electrons are taken by the sodium ions from the cathode. We see therefore that so far as the battery and lead wires are concerned a current flows in the circuit. Within the solution the conduction process, which is really an electro-chemical conversion, is complicated compared with the conduction in metallic wires. Conduction within the solution consists of a slow migration of positive and negative ions.

Semiconductors

Between the two extremes of conductors and insulators there is a class of materials known as semiconductors. As the name suggests, the materials are neither conductors or insulators, but sufficient conduction occurs to make these materials useful. Copper oxide, selenium, gallium arsenide, and the sulphides of lead and cadmium are semiconductors, but the two best known materials are **germanium** and **silicon** because transistors are made from these elements.

Germanium and silicon are both **tetravalent** elements. Atoms of these elements bond together by the sharing of electrons in the outer shell. Four electrons from any one atom are shared with four atoms in the immediate vicinity. When electrons are shared they form what is called a **covalent bond**, which is usually a strong binding force. (Consider, for example, how strong and hard a diamond crystal is.) Silicon and germanium atoms bind together in a crystal form which is similar to that of the diamond lattice. Since each atom shares four electrons, four covalent bonds are formed, hence the term tetravalent (tetra = 4). When techniques are used to produce transistor material, extremely pure single crystals are formed. The aim of modern technology is to produce mechanically perfect crystals in which the impurity content is no greater than one part in 100,000,000.

A mechanically perfect crystal at absolute zero (about −273°C) is a perfect insulator because all of the covalent bonds are properly formed, and no free charge carriers exist within the crystal. At room temperatures, however, germanium and silicon crystals, being semiconductors, contain some charge carriers. This is because there is always a possibility that at any point in the crystal a covalent bond is ruptured. The energy to break down a covalent bond usually comes from heat sources, but alternative sources of energy are light energy or radiation energy from, say, some radioactive source.

When a covalent bond is broken an electron is released into the crystal. The absence of an electron is in effect a positive charge since the parent atom has lost a negative electron. The absence of an electron is called a **hole**. It is convenient to regard a hole as a separate entity having a positive charge equal in magnitude or size to the negative charge on the electron. An analogous concept arises when we refer to the absence of water in a vertical tube as a bubble. As the bubble floats up the tube we say that the bubble has moved, although we realize that a more complicated movement of water molecules is involved. In the same way, when an electron drops into a hole (and the valence bond is re-established) the region from which the electron came must be positively charged. It is convenient to regard this as the movement of a hole from the position where the covalent bond is reformed to the position from where the electron came. At room temperatures, therefore, within a crystal of say silicon, many covalent bonds are broken. Electrons are ejected into the crystal, leaving behind holes. The holes are not fixed in position, however, because any electron from some other part of the crystal may re-establish a covalent bond. The region from which the electron came is a hole, hence we may regard the original hole as moving from its first position into another part of the crystal. When an electric field is established in the crystal—say, by connecting a voltage across the crystal—the charge carriers move; the electrons are urged to the source of positive potential, whilst the holes are urged towards the negative potential source. Conduction therefore takes place in the crystal by the movement of two types of charge carriers; such conduction is termed **intrinsic conduction**, and in so far as it takes place it is a nuisance. If intrinsic conduction were absent semiconductor devices would be superior to those available today.

Since silicon has a much lower intrinsic conduction than germanium, most transistors and allied semiconductor devices are made from silicon.

Before the discovery of the electron in 1897 it was generally assumed that electric current flowed from a positive potential region to a negative one. When it was realized that for conduction in metals and gases the current consisted of electrons (and hence travels in the opposite direction to that previously supposed), some consideration was given to reversing the previous convention. Such a major upheaval in the reprinting of books, articles and circuit diagrams would have resulted that the idea was abandoned. We see that in recent times the idea of the direction of 'conventional current' flow from positive to negative need not upset us when we consider semiconductors in which holes exist.

Impurity Conduction

One way of increasing the conductivity of a crystal is to add a controlled amount of certain impurities. The conduction that then occurs is called **impurity conduction** and is of paramount importance in the operation of semiconductor devices.

When small amounts (1 part in 1 to 10 million) of a pentavalent impurity such as arsenic or phosphorus are added during crystal formation, the impurity atoms lock into the crystal because the size of such impurity atoms is not greatly different from that of the silicon atom. The crystal lattice is not therefore unduly distorted. Now a **pentavalent** impurity atom, as the name suggests, has five valence electrons. Four of these form the normal covalent bonding with adjacent silicon atoms, but the fifth electron has no proper valence role to play. Such an electron is very easy to detach from the parent impurity atom, and at normal room temperatures, say 20°C, these electrons become free and wander within the crystal with random motion. The pentavalent impurity donates an electron to the crystal and is therefore called a **donor impurity**. Because of the presence of free negative charge carriers (the electrons), the conductivity of the crystal is increased and the material is then known as an *n*-type silicon. The crystal as a whole is, of course, still electrically neutral because there are as many positively charged donor centres (impurity atoms) as there are negative electrons. But the ionized impurity atoms are fixed in position so, not being able to move, they play no part in the conduction process.

Instead of adding a pentavalent impurity to an otherwise pure crystal, a **trivalent** impurity such as boron may be added. (The jargon term for the addition of an impurity is called 'doping' by semiconductor manufacturers.) The three valence electrons enter into normal covalent bonding, but a region of stress exists where the fourth bond ought to be. Any stray electron that happens to be in the vicinity can be trapped by the impurity centre, which then becomes negatively charged. The stray electron will have come from a region in the crystal where a covalent bond has been broken. Remembering that as many holes as free electrons are created when a covalent bond is ruptured, once such electrons are trapped in the trivalent impurity (called an **acceptor impurity**) a surplus of holes exists within the crystal. In effect, therefore, the incorporation of trivalent impurities into an otherwise pure silicon crystal has the effect of releasing many holes within the crystal. Such holes, being positive charge carriers, are available for conduction, and the material is then known as **p-type silicon**. Once again, however, it should be realized that the crystal as a whole is electrically neutral. The representation of *n*-type and *p*-type silicon is shown in Fig. 23.

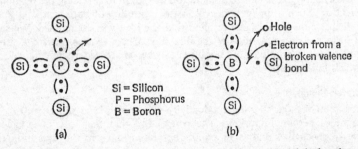

Si = Silicon
P = Phosphorus
B = Boron

(a)　　　　　　　　　　　　　　　(b)

Fig. 23. (a) Representation of *n*-type silicon. Conduction is mainly by electrons because they are in the majority; some holes are present because of the breakdown of valence bonds, but they are in the minority. (b) Representation of *p*-type silicon. At room temperatures sufficient electrons are present from the breakdown of valence bonds to fill the acceptor centres. The region from which the electron came is a 'hole'. Holes are the majority carriers in *p*-type silicon.

In an *n*-type semiconductor, electrons are in the majority and are called **majority carriers**. Some holes also exist owing to the formation of electron-hole pairs at room temperatures; these are **minority carriers**. Alternatively, in a *p*-type semiconductor, holes are the majority carriers and electrons are minority carriers. The number of charge carriers present determines the conductivity of the crystal.

The *pn* Junction Rectifier

If a piece of *n*-type silicon is joined to a piece of *p*-type silicon an important electronic device results. We shall see that such a device can easily pass current when a steady applied voltage is connected one way, but when we reverse the polarity of the supply voltage practically no current flows. We have made a **diode**, or **rectifier**. The device is called a diode because two electrode regions are involved; alternatively we may call the device a rectifier because if alternating voltages are applied across it, current flows only during the time that the polarity of the supply is such that current flows easily. This device is therefore the basis of circuitry that can convert alternating voltages into direct or steady voltages, a process known as **rectification**.

We cannot, of course, make a diode by sticking together with glue a piece of *n*-type silicon and a piece of *p*-type silicon. The discontinuity at the boundary of the two crystals would be so severe that rectifying action could not take place; indeed no electrons could cross such a barrier. The *p*- and *n*-type silicon must be joined in such a way that both pieces form part of a single continuous crystal structure. The junction between the two types of semiconductor is then known as a *pn* junction. Several ways have been tried in the past to produce such a junction, and some of them have been successful. Present-day commercial methods rely on the process of solid-state diffusion for forming *pn* junctions.

The semiconductor material most generally used is **silicon**. The first step in the manufacture is the growth of a single crystal of silicon. Very pure silicon, together with small accurately known amounts of impurity are placed in a quartz crucible set into a graphite receptacle. The contents of the crucible are then heated in an inert atmosphere, by means of radio-frequency heating coils, until completely molten. When the temperature of the melt is just above the melting point of silicon a seed crystal, which is a small perfect single crystal of silicon, is dipped into the surface of the melt and slowly withdrawn. Provided the correct techniques are used a single crystal ingot some 40 mm in diameter can be drawn out of the melt. This will usually be *p*-type silicon because later stages in the production of semiconductor devices, especially integrated circuits, are made easier.

The ingot, after undergoing a further purification process, known as zone refining, is cut into thin slices by a special diamond saw. After grinding and polishing the slices to as near a flawless finish as possible, they are subsequently washed and etched in acid. The slices are then placed in a diffusion furnace and heated to temperatures in the region of 1000°C to 1200°C. This is below the melting point of silicon, so the slices do not melt; at this high temperature, however, diffusion processes are greatly accelerated. The surface of the *p*-type slice is then exposed to a vapour containing a pentavalent impurity such as arsenic or phosphorus. Provided the conditions are correct the pentavalent impurity diffuses into the slice, converting only the uppermost thin layer of *p*-type silicon to *n*-type silicon; the *p*-type silicon remains underneath. Within a single crystal structure therefore we have a *p*-type silicon layer joined to an *n*-type silicon layer. Such a *pn* junction is found to be an extremely efficient rectifier mainly because such layers are so thin that losses within the bulk of the crystal are minimized.

To prevent contamination and deterioration of the surface a layer of silicon dioxide, which is an extremely good insulator, is grown epitaxially (see below) on the surface by exposing the latter to steam and oxygen at high temperatures. Part of the protective silicon dioxide layer is subsequently etched away with hydrofluoric acid, and a metal contact is made to the underlying *n*-type layer. Another contact is made to the *p*-type substrate beneath.

Many hundreds of diodes are fabricated simultaneously on one slice. Subsequently the diodes are separated by cutting the slice up with a diamond scorer; the leads are then added and the device encapsulated in a metal or plastic container. Fig. 24 shows diagrammatically the geometry of such a junction.

Mention has been made of the fact that the insulating layer of silicon diode is grown epitaxially. The technique of epitaxial growth together with the diffusion process is the standard way of making modern diodes, transistors and integrated circuits. The word *epitaxial* comes from the Greek (*epi* = upon and *taxos* = arranged). It is essential for correct diode and transistor operation that the various *p*- and *n*-type regions exist in a single crystal structure. Epitaxial growth is a way of ensuring this. By exposing the surface of the slice to various vapours, in turn, layers of silicon, silicon dioxide, *p*-type or *n*-type silicon can be made to grow on the surface in such a way that the single crystal structure is preserved, i.e. we grow various layers upon (*epi*) the crystal arrangement (*taxis*). Because epitaxial layers are grown on flat surfaces, or planes, the diodes and transistors fabricated by this technique are called **planar epitaxial** devices.

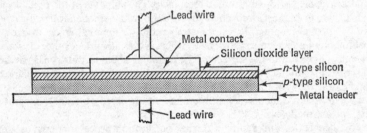

Fig. 24. A planar epitaxial *pn* junction rectifier.

The Rectification Process

Bearing in mind that the true geometry of a rectifier is something like that shown in Fig. 24, we can see how a *pn* junction rectifier works by considering the 'bar' type diagram of Fig. 25. Upon formation of the *pn* junction there is initially on each side of the junction a number of free electrons in the *n*-type silicon and a number of free holes in the *p*-type silicon. Because the crystal structure is preserved across the junction, electrons from the pentavalent impurity centres diffuse across the junction and are trapped in the acceptor impurity centres in the *p*-type silicon. The result is that a layer of negative charge lies on the *p*-type side of the junction. When the electrons from the pentavalent impurity centres diffuse away to the *p*-type material, a layer of positive charge lies on the *n*-type side of the junction. This is because each donor centre, in giving up an electron to the crystal (and thus losing a negative charge), is left with a net positive charge. As more and more electrons diffuse across the boundary the negative and positive layers increase in strength until eventually no more electrons can diffuse across the junction because of the repelling effect of the sheet of negative charge on the *p*-side of the junction. Similarly, holes in the *p*-side are prevented from diffusing across the junction. The result is that at the junction we have a region which is depleted of charge carriers. This region is sometimes called a **depletion layer**. Because there are no charge carriers in the depletion layer such a layer becomes an insulator and constitutes a barrier to the passage of electric current.

When a source of voltage is applied with the positive terminal connected to the *n*-type side of the crystal and the negative terminal connected to the *p*-type side, the depletion layer becomes thicker. This is because the positive terminal of the voltage source attracts electrons away from the junction on the *n*-type

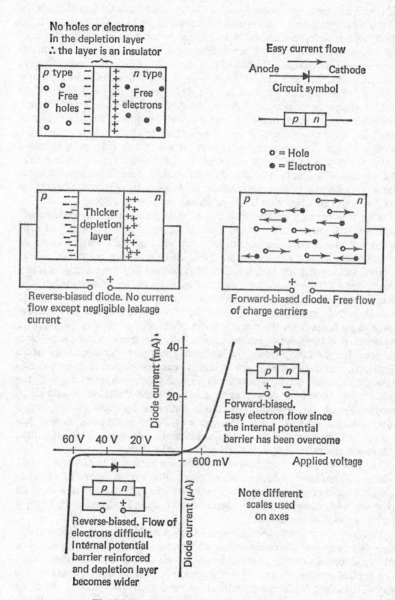

Fig. 25. Rectifying action of a junction diode.

side, and the negative terminal attracts holes away from the junction on the p-type side. Attracting electrons away from a sheet of positive charge is, in effect, adding holes to the sheet, thus increasing the strength and thickness of the charge. A corresponding mechanism operates on the p-side, where the sheet of negative charge is thickened because holes are attracted away from it. Since the depletion layer is now thicker, the region is an even better insulator. As long as the insulator exists a current cannot flow. When the polarity of the voltage source is as described above the diode is said to be **reverse-biased**. The extremely small current that does flow can be ignored for practical purposes. Such a current results from the formation of electron-hole pairs in the depletion layer. The holes are swiftly swept across to the p-side, being attracted by the negative layer of charge. The electrons are similarly swept to the n-side. We see now why silicon is preferred to germanium. Silicon has a much lower intrinsic conductivity than germanium, and hence fewer electron-hole pairs are formed in the depletion layers of diodes made from silicon.

When the polarity of the voltage source is reversed the depletion layer becomes narrower as the voltage is increased from zero up to about 600–700 mV. The current is therefore slow to rise as the applied voltage is increased up to these values because a depletion layer still exists. Once the applied voltage is increased beyond 700 mV, however, it is noticed that the current increases enormously. This is because the negative terminal of the voltage source, now being connected to the n-side, repels the electrons with sufficient force to enable them to penetrate the depletion layer. At the same time an attractive force urges holes from the positive sheet of charge to move to the negative terminal. A corresponding attraction exists for electrons forming a negative sheet of charge on the p-side of the junction. The sheets of charge are therefore dispersed, and very little resistance exists once the depletion layer has disappeared. Current can now flow freely through the crystal structure. With the polarity of the voltage source applied in this sense the diode is said to be **forward-biased**. A forward-biased diode therefore conducts current, but a reverse-biased diode does not. The action of diodes in circuits can be understood if it is remembered that a forward-biased diode acts as a closed switch, whereas a reverse-biased diode acts as an open switch.

The most important application of diodes is their use in power supplies, where it is necessary to convert alternating voltages from the mains to steady voltages for energizing electronic circuits. As this is the subject of a separate chapter, no discussion will be undertaken here, except to explain the elementary principles of rectification.

Fig. 26 shows a circuit of a diode and a series load resistor R_L, the combination being supplied with a sinusoidal alternating voltage. This voltage may be obtained from an oscillator or the secondary winding of a transformer. When the applied voltage is such that the anode is positive with respect to the cathode, current flows in the circuit. If we neglect the resistance of the diode then the current waveform will be sinusoidal during the first half cycle. When the polarity of the voltage is reversed, the anode becomes negative with respect to the cathode and no conduction takes place. Fig. 26 shows the position in graphical form. The voltage across the resistor is proportional to the current and therefore this voltage will also have a sinusoidal waveform for each positive half cycle. The alternating voltage has thus been converted into a series of unidirectional pulses, a process known as **rectification**.

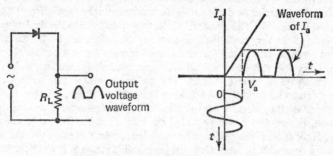

Fig. 26. Basic principles of rectification.

Point-Contact Diodes

Historically, point-contact diodes were in use long before *pn* junction diodes. Many readers will know of the cat's whisker era in the 1920s. Radio was becoming increasingly popular and many amateurs built their own crystal receiving sets. In order to demodulate the radio waves it was necessary to use a metal-to-semiconductor rectifying diode. As has already been mentioned earlier in this chapter, the crystal involved was a lead sulphide crystal (impure galena). The sharp point of a brass wire was brought into contact with the crystal, the contact being maintained by bending part of the wire into a spring—hence the name cat's whisker.

Although such rectifiers are not used today it is still necessary to use the principle involved when radio signals of high frequency have to be detected. The ordinary *pn* junction has quite a large capacitance associated with it when reverse-biased. This is because an insulator, the depletion layer, is separated by two conductors, the *p*- and *n*-type silicon. Charge is stored in such a capacitor. The capacitance turns out to be too large for the detection of very high frequency signals. At such high frequencies the reactance of even a small capacitor can be quite low, perhaps only tens or hundreds of ohms. This means that during the half cycle when no current should flow an appreciable charging current flows. The rectification properties are therefore largely nullified.

The difficulty can be overcome by using a metal-to-semiconductor diode. Such an arrangement involves holding the end of a metal wire in contact with *p*-type or *n*-type silicon. A diode of this type cannot store charge, and in any case the effective rectifying area is very small. No disturbing capacitative effects are therefore evident even at high frequencies.

Because of the small rectifying area involved, the current-carrying capacity of a point-contact diode is small. Whilst this is of no consequence when using the device as a signal diode, such a limitation would be severe when rectifying large alternating currents. For power supplies therefore, *pn* junctions are needed.

Zener Diodes

When a *pn* junction diode is biased in the reverse direction the majority carriers (holes in the *p*-side and electrons in the *n*-side) move away from the junction. The barrier or depletion layer becomes thicker and current transfer is almost impossible. The presence of electron-hole pairs in the depletion layer accounts for the very small leakage current that can be detected. This leakage current remains very small for all reverse voltages up to a certain threshold value. Once this threshold voltage has been exceeded there is a sudden and substantial rise in reverse current. The voltage at which this sudden rise in current occurs is called the **avalanche** or **breakdown voltage**, or often the **Zener voltage**, after the scientist who studied this phenomenon. The breakdown, however, is non-destructive provided the current is insufficient to cause the power dissipation of the device to be exceeded; under this condition the breakdown is a reversible process. Over the permissible range of reverse current it is found that the voltage across the diode remains nearly constant, as is evident from observing the characteristic curve, a typical example of which is shown in Fig. 27. For this reason the most common application of zener diodes is in voltage regulation or stabilization.

We often find in electronic circuitry that supply voltages vary. There are several reasons for this; perhaps in mains-operated equipment the mains voltage varies; in battery-operated equipment the battery voltage diminishes with age and use; or the supply voltage varies because the load current varies. One way of stabilizing the supply voltage is to use the simplest circuit shown in Fig. 27. The design procedure for a simple stabilizer of the type shown is as follows.

(*a*) Determine the maximum and minimum values of load current, I_L. (Even if the minimum value of load current is considerable it may well be a good idea to assume zero value, since the load may well be inadvertently disconnected.)

(*b*) Determine the minimum value that the supply voltage is ever likely to be. In any case, for proper operation the supply voltage must always be at least one volt (and preferably two or three volts) higher than the breakdown voltage of the diode.

(*c*) Select a suitable zener diode. Zener diodes are manufactured in a range of preferred values (e.g. 4·7 V, 6·8 V, 18 V, 39 V, 47 V, etc.). (It is possible to add two or more zener diodes in series to make up an intermediate value. The current passing through the series must not exceed the maximum current rating for any one diode.) Three quantities need to be known about the device: the breakdown voltage, the wattage rating and the tolerance (10, 5 and 1 per cent are typical tolerance figures). The wattage rating must be greater than $V_z I_{z(max)}$, where V_z is the breakdown voltage and $I_{z(max)}$ is the greatest current the zener diode is called upon to conduct.

(*d*) Calculate the resistance value and wattage dissipation of the resistor R_1 (Fig. 27).

The stabilizing action is then as follows: the voltage across the load R_L will be reasonably constant and equal to the zener voltage V_z. We say 'reasonably constant' because some small variation of V_z is noticeable when the current through the zener diode varies. For a given *fixed* supply voltage, V_1, the voltage across the resistor R_1 will be $V_1 - V_z$. Since this is constant, the current through R_1 must be constant. This means that the sum of the zener current and the current through the load must be constant. If the load current through

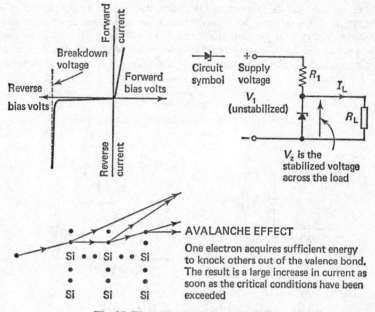

Fig. 27. The zener diode characteristics.

R_L increases, there must therefore be a corresponding fall of current through the zener diode. Conversely, if the current through R_L falls, the current through the zener diode must rise by a corresponding amount. We see therefore that if the current through R_L falls from its maximum amount to zero, the current through the zener diode will rise by this amount; the diode must therefore be capable of conducting with safety the maximum load current.

Another important point to note is that when the load current rises to a maximum the diode current must fall, but we must not allow this to fall to zero. If we did we would lose the voltage control. This can be seen by examining the zener characteristic. As the zener current falls we eventually reach the knee of the reverse characteristic. At this point a current of 1 or 2 mA will be flowing. If we reduce the zener current below this value we pass the knee and approach the origin, i.e. the voltage across the diode falls from the zener breakdown voltage to zero. Consider now the position when the supply voltage, V_1, instead of being fixed, rises. This will cause the supply current to rise. The voltage drop across R_1 then increases by an amount equal to the rise in supply voltage; the voltage across R_L therefore remains the same.

Let us illustrate the design procedure by considering an example. Suppose we wish to operate a piece of equipment at a constant voltage of 12 V, and that the equipment may draw 100 mA as a maximum current; a supply voltage of 15 V (nominal) is available, and it is known that this voltage can never exceed 16 V. The problem is to design a suitable stabilizer circuit. We would proceed, as explained above, to find a suitable resistor for R_1, and a suitable zener diode. After noting that the maximum load current is 100 mA, and assuming a minimum value of zero, we conclude that the maximum zener current is 102 mA (100 mA for the load plus 2 mA to ensure that we do not move past the knee of the characteristic). Obviously we would select a diode with a zener voltage of 12 V. The maximum power dissipation of the diode is therefore

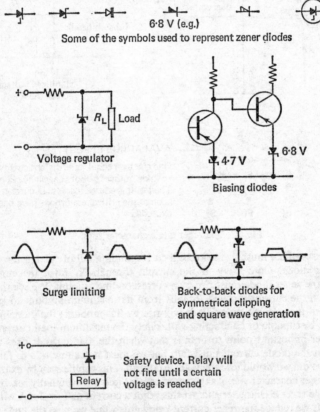

6·8 V (e.g.)

Some of the symbols used to represent zener diodes

R_L Load

Voltage regulator

Biasing diodes

4·7 V   6·8 V

Surge limiting

Back-to-back diodes for symmetrical clipping and square wave generation

Safety device. Relay will not fire until a certain voltage is reached

Relay

Fig. 28. Some applications of zener diodes.

$0.102 \times 12 = 1.224$ watts, i.e. approximately 1·25 W. A standard commercially available 1·3 watt diode would then be adequate for our purposes, and we would have a safety margin in hand. Since the maximum supply voltage is 16 V the maximum voltage across the resistor R_1 must be $16 - 12 = 4$ V. The maximum current is 102 mA; therefore $R_1 = 4 \div 0.102 = 40$ ohms. (Readers should become increasingly skilled at making arithmetic approximations. Bearing in mind that a resistor with a tolerance of ± 10 per cent is likely to be used, and that in any case the nearest preferred value must be selected in the majority of cases, it is not sensible to quote figures to several decimal places.) To calculate the power rating of R_1 we note that $4 \text{ V} \times 0.1 \text{ A} = 0.4 \text{ W}$; therefore a half-watt resistor would be adequate.

Some of the applications of zener diodes are shown in Fig. 28.

The Junction Transistor

The first transistors emerged from an American company (the Bell Telephone Laboratories) in 1948. The American physicists Bardeen and Brattain were investigating the behaviour of point-contact diodes by probing the area near to the contact point with a second metal wire. Their object was to try to improve their understanding of the fundamental physics involved. Almost by accident they discovered that when probing the base semiconductor material, the current through the probes (one emitting current into the base material and the other collecting it from the base) was not only much greater than the current through the base material from the emitter probe, but that the two currents were proportional over a useful range. Unlike a rectifier. therefore, this new arrangement could be made to increase, i.e. amplify, currents. Prior to 1948 the only electronic amplifying devices known were thermionic triodes and pentodes, along with their variants. It is difficult to find any other invention that surpasses in importance the transistor; this device has revolutionized the whole of technology during the last twenty-five years.

It was found that the point-contact transistor was not easy to reproduce in large numbers by mass production techniques. Fortunately a much more practicable device was invented by Bardeen and Brattain's co-worker, Shockley, in 1950, and the early 1950s saw the limited production of junction transistors.

The **junction transistor** is a sandwich arrangement in which a system of two closely-spaced *pn* junctions is formed within a single crystal structure. The transistor may be constructed in several different ways. The early devices were made by an alloyed junction technique in which two pellets of indium were alloyed to an *n*-type germanium wafer some tenth of an inch square, one on each side. During subsequent heat treatment the trivalent indium diffused into the *n*-type base material, converting some of the base material to *p*-type material. The diffusion was stopped just before the *p*-zones met. An extremely thin section of *n*-type germanium is thus sandwiched between two layers of *p*-type germanium. A *pnp* transistor was the result. It was found that the frequency response (i.e. the range of frequencies over which the device would act as an amplifier) and the power handling capability were limited. The frequency response is limited by the time it takes for the charge carriers to cross the base region. The technique of making alloyed transistors does not allow bases to be made much thinner than about 25 microns (about 1 thousandth of an inch). Although much effort was expended in improving the technique, the methods have been superseded by the invention of the planar epitaxial technique. This process leads to the production of the most versatile and reliable transistors presently available, and it is difficult to see any process supplanting the epitaxial method for many years.

Silicon was chosen rather than germanium because several advantages were obtained. Silicon transistors can operate at higher temperatures than their germanium counterparts, can be operated at higher voltages and have much smaller leakage currents. The technology of silicon was developed to such a state that mass production techniques could be used.

We have already described the planar epitaxial technique for producing diodes. The principles of producing transistors is similar. Fig. 29 shows diagrammatically a typical planar epitaxial transistor. A low-resistivity single-

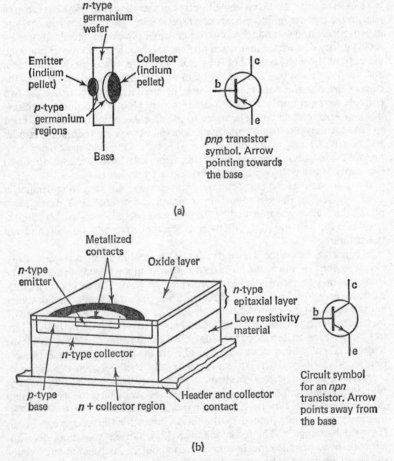

Fig. 29. Constructional features of different types of transistor. The planar epitaxial type is the standard type used now. (a) Alloy transistor; (b) planar epitaxial transistor.

crystal substrate acts as the collector region. The low resistivity is achieved by doping this part of the crystal heavily. Electrical losses in the bulk of the collector region are then low owing to the small resistance. However, in order to be able to operate the transistor with voltages as high as say 50 V or more, the base-to-collector junction must be formed in high-resistivity material. The manufacturers therefore grow an extremely thin layer of almost pure silicon upon the heavily doped substrate; this is the epitaxial layer within which the transistor is fabricated. Thus is produced a device which has low collector losses yet reasonably high operating voltages. The base and emitter regions of the transistor are then produced by diffusing into the epitaxial layers the necessary impurities. The correct geometry is achieved by first growing a layer of silicon dioxide on the surface and then etching away the required window by a photoengraving process. Impurities are then allowed to diffuse downwards and underneath the oxide layer to form the base and emitter regions. Suitable contacts are then made to the regions. The electrical performance of devices made in this way is excellent.

Transistor Operation

Since a transistor consists of two *pn* junctions within a single crystal, transistor action can be explained with reference to Fig. 30. For diagrammatic purposes the base region is shown fairly thick, but in fact the *pn* junctions are very closely spaced and the active portion of the base is very thin. In the absence of

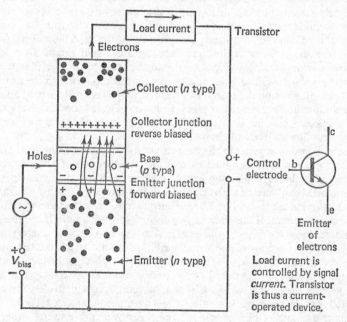

Fig. 30. Diagrammatic representation of the amplifying action of a transistor.

any external applied voltages, the collector and emitter depletion layers are about the same thickness, the widths depending upon the relative doping of the collector, emitter and base regions. During normal transistor operation the emitter-base junction is forward-biased so that current flows easily in the input or signal circuit. The bias voltage is about 200 mV for germanium transistors and about 650 mV for silicon devices. (Fig. 30 shows the bias voltage being obtained from a battery. In practice, as we shall see in the chapter on amplifiers, this bias is obtained in other ways.) The collector-base junction is reverse-biased by the main supply voltage. Voltages of 6 V, 9 V and 12 V are common. The collector junction is therefore heavily reverse-biased and the depletion layer there is quite thick. The voltage gradient in the depletion layer is high because nearly all of the battery voltage appears across it. There is little loss of voltage in the bulk of the collector portion of the crystal because most of this bulk is made from low-resistivity material. In epitaxial transistors this is achieved by using heavily doped material; in alloyed types the collector pellet is considerably larger than the emitter pellet. The voltage drop across the emitter junction is low and can be ignored.

In the absence of a signal, the bias source injects a steady stream of holes into the base region. The entry of every hole into the base region is accompanied by the entry into the base of an electron from the emitter region. Combination of the hole and the electron is not likely to occur, however. The base region is lightly doped compared with the emitter region and so the lifetime of the electron in the base region is quite long. In addition, the base is extremely thin, so the electron, instead of combining with a hole, is much more likely to diffuse into the base-collector junction. The electron then comes under the influence of the strong electric field in the depletion layer and is swept into the collector region and hence into the load circuit. In a good transistor many electrons pass into the load circuit before the hole which entered via the base lead is eliminated on combination with an electron. A small current entering the base thus gives rise to a large load current, and so current amplification has taken place. In practical transistors, for every hole that is injected into the base some 50 to 400 or more electrons may be influenced to flow in the load circuit.

The presence of a signal in the base circuit modulates the number of holes injected into the base. The load current therefore consists of a steady component upon which is superimposed the amplified version of the signal current. The importance of having a bias current will be appreciated when it is realized that the varying component of the load current must have the same waveform as the signal current. If this were not so, distortion would result. It is therefore useless to have the signal voltage alone in the base-emitter circuit since negative-going portions of the signal waveform would cut off the transistor and rectification would result. (When the voltage applied to the base is such that the load current is reduced to zero, the transistor is said to be cut off.)

A *pnp* transistor behaves in a similar fashion except that electrons are injected into the base and holes flow from the emitter into the collector. To maintain the correct bias conditions the polarity of the external batteries must be reversed.

When using a transistor in the way described above it will be noticed that the emitter lead is common to both the signal circuit and the load circuit. Operation is then said to be in the **common-emitter mode**. In the USA this is referred to as the **grounded-emitter mode** because the emitter is connected to the grounded (i.e. earthed or chassis) line. This is the most frequently used voltage amplifier arrangement because, compared with the common-base mode described below, little current is drawn from the signal source. An alternative way of viewing this is to say that the input resistance of the common-emitter amplifier is higher than that of the common-base mode.

There is a range of input base currents over which the load current is directly proportional to the base current. This is the operating range. The change in collector current, i_c, divided by the change in base current, i_b, is called the **current gain** of the transistor and is given the symbol h_{fe}. (α' or β is used in older literature.) The change in base current is of course the **signal current**. Since the load current is controlled by a signal current the transistor is regarded as a current-operated device. The collector current in a transistor does, of course, depend upon the base-emitter voltage, but the relationship is non-linear.

Fig. 31 shows the three basic transistor arrangements. The **common-emitter mode** is the most commonly used arrangement for voltage amplification because, as has already been pointed out, very little current is required from the signal source.

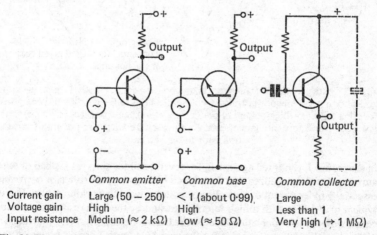

	Common emitter	Common base	Common collector
Current gain	Large (50 – 250)	< 1 (about 0·99)	Large
Voltage gain	High	High	Less than 1
Input resistance	Medium (≈ 2 kΩ)	Low (≈ 50 Ω)	Very high (→ 1 MΩ)

Fig. 31. The three basic amplifier arrangements together with some of their properties. The figures are given only as a guide to the magnitudes involved.

The **common-base** mode of operation is also capable of voltage amplification in spite of the fact that in this mode the collector current is less than the emitter current by an amount equal to the base current. For this mode the current gain is therefore less than unity. Voltage amplification is achieved by the use of high values of load resistor. The transistor is able to maintain the current through the load because the device is a good constant current generator; that is to say, it can deliver a current that is substantially independent of the load resistor provided variations of this resistor fall within reasonable limits.

Since the circuits shown in Fig. 31 are to be discussed in detail in the chapter on amplifiers nothing further need be said here.

A typical set of characteristic curves for the common-emitter mode is shown in Fig. 32. The characteristics for the common-base mode are similar

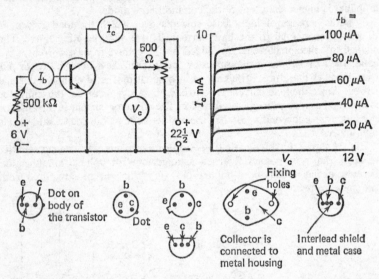

Fig. 32. Characteristics of a transistor in the common-emitter mode. Figures given are typical of a small low-power transistor. Modifications to the circuit will be required for other types (e.g. high-voltage or power transistors). Some connection diagrams are given for transistors in common use. In all cases the wires are pointing from the transistor to the viewer.

in shape, but I_c is plotted against V_c for several emitter currents instead of base currents. It will be realized that for a given base current the collector current is essentially independent of the collector voltage over the working range. This is typical of a device with a high internal resistance; the high resistance is associated with the wide depletion layer at the base-collector interface. With such a high internal resistance, variations of external load resistance do not affect the collector current very much.

The Unijunction Transistor

This type of transistor consists of an *n*-type silicon bar with ohmic contacts (called base one, b_1, and base two, b_2) at opposite ends. (An ohmic contact is one that conducts currently easily irrespective of the direction of flow of the current.) A single rectifying contact is made between the two base contacts, usually closer to b_2 than b_1. This rectifying contact is called the **emitter**. During normal operation b_2 is held positive with respect to b_1 via a suitable load resistor (Fig. 33). Provided the emitter junction is reverse-biased, there is no emitter current and the *n*-type silicon bar acts as a simple potential divider, the voltage at the emitter contact being a fraction, β, of the voltage from b_2 to b_1. In order to maintain this condition the emitter junction must remain reverse-biased; this is ensured by keeping the emitter voltage, V_e, less than

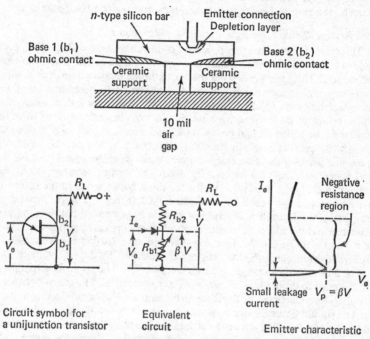

Fig. 33. The unijunction transistor of the General Electric Company.

βV, where V is the voltage from b_2 to b_1. When V_e is allowed to exceed βV the junction becomes forward-biased and current flows in the emitter-to-b_1 region. The emitter injects holes into this region, and as they move down the bar there is an increase in the number of electrons in the lower part of the bar. The result is a substantial decrease in the bar resistance from the emitter to b_1. This decrease in emitter-to-b_1 resistance brings about a fall of emitter voltage. A condition arises whereby increases of emitter current are accompanied by decreases of emitter voltage, resulting in a negative resistance characteristic.

Unijunction transistors are low in cost and are characterized by a stable triggering voltage, which is a fixed fraction of the voltage between b_1 and b_2. They have a negative resistance characteristic which is uniform among units of the same type and which is stable with temperature and throughout the unit's life. Apart from its use in oscillators and time-delay circuits, the major use of a unijunction is as a pulse generator to fire silicon-controlled rectifiers. This application is the subject of a later chapter.

Field-Effect Transistors

Although it has been possible for more than thirty years to make field-effect devices in the laboratory, it is only recently that the significant advances in semiconductor technology have made possible the manufacture of reliable devices in large numbers. Basically there are two types of field-effect transistor, namely, the reverse-biased *pn* junction type and the insulated-gate device.

The Reverse-Biased Field-Effect Transistor (Junction F.E.T.)

This type of device was first proposed by Shockley, who called it a unipolar field-effect transistor because only one type of charge carrier is used to carry the current. This is different from the conventional bipolar transistor in which two types of charge carrier are involved. It will be remembered that for an *npn* transistor holes are injected into the base and these holes induce many electrons to flow in the collector or load circuit. The current that flows from the emitter to the collector, and hence to the load circuit, is controlled by a much smaller current that enters by the control electrode, i.e. the base. A field-effect transistor works quite differently. Current flows from the source to the drain and hence into the load circuit. The source of an f.e.t. corresponds to the emitter of a bipolar transistor, whilst the drain corresponds to the collector. This current is controlled by an electric field (hence the term 'field-effect' transistor), which itself is determined by a voltage applied between the gate electrode and the source. The gate electrode therefore corresponds to the base of a bipolar transistor. An f.e.t. can therefore be regarded as a voltage-controlled device because the load current is controlled by the gate voltage.

Fig. 34 shows the cross-section of a junction f.e.t. The control current flows from source to drain along a channel of *n*-type material. The majority carriers are therefore electrons in this case. Although an *n*-channel f.e.t. is shown, readers should be aware that *p*-channel versions are also readily available. (In the *p*-channel f.e.t. the majority carriers are holes. Although the principle of operation of both types is the same, we shall consider the *n*-type version; the description applies to *p*-channel types, but the polarity of the applied voltages must be reversed.)

The *n*-channel is sandwiched between two *p*-regions. These regions are connected together electrically and are biased negatively with respect to the channel. This is achieved by making the gate negative with respect to the source. The *pn* junctions at the gate-channel interface are hence reverse-biased and therefore a depletion layer extends into the channel. The shape of the depletion layer is not uniform, however, and depends upon the voltage applied between the drain and the source (V_{DS}). For voltages greater than 1 or 2 V the depletion layer becomes wider as the drain region is approached. This is because the reverse-voltage increases as the drain region is approached. For drain-to-source voltages of approximately any given gate voltage and above, the two depletion layers converge and the channel approaches what is known as the **pinch-off** condition. As the depletion layers become closer and closer

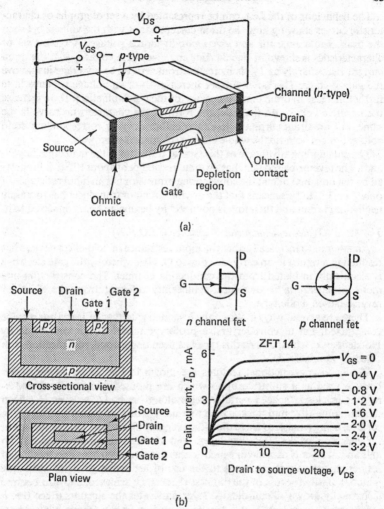

Fig. 34. (a) The Shockley-type field-effect device. (b) A modern form of construction due to Ferranti Ltd. For many applications gate 1 and gate 2 are connected together in the external circuit.

with increasing gate and/or drain voltages the resistance of the channel rises. It might be expected that the channel resistance becomes infinite as the depletion layers from each side meet, thus cutting off the drain current. This does not happen, however, owing to the screening effect of the charge carrier density at the centre of the channel. We find that once this condition has been reached the drain current is almost independent of the drain voltage, and a saturation effect is observed.

The behaviour of the f.e.t. can be represented by a set of graphs or characteristic curves showing how the drain current varies with the voltage between the drain and source for any given gate-to-source voltage. A typical set of characteristics is shown in Fig. 34. For any given gate voltage, V_{GS}, the drain current rises sharply as V_{DS} is increased from zero up to a voltage just above the gate voltage. Thereafter further increases in drain voltage produce little further increase in drain current because of the pinch-off effect. If we increase the negative voltage on the gate, the form of the characteristic curve is the same, but the drain current is less than before at every corresponding drain voltage; this is because the whole effective channel is narrower.

The outstanding advantage of this type of transistor is the high input impedance. The reverse-biased f.e.t. has an input resistance of about 10^{10} (i.e. 1 followed by ten noughts) ohms; the conventional transistor has an input resistance of only 2 or 3 kΩ. This means that the driving circuit or transducer has to supply negligible current and therefore is not upset by having an f.e.t. connected to it.

The Metal-Oxide-Semiconductor Transistor (M.O.S.T.)

An attempt to increase further the input resistance of field-effect devices has resulted in a return to an early idea due to O. Heil, whereby the gate electrode is electrically insulated from the conducting channel. The construction and mode of operation is therefore significantly different from the Shockley reverse-biased diode type.

There are several ways in which the insulated gate type of transistor may be constructed. Let us consider first a prototype model of the kind made by Hofstein and Heiman. Once this type has been understood, modifications can be easily appreciated.

The main constructional features are shown in Fig. 35. A p-type silicon body is used as a substrate upon which are diffused two heavily doped n-regions in closely spaced parallel strips along the body. A layer of silicon dioxide some 10^{-4} mm thick is then thermally grown or evaporated on to the surface, using a mask to leave the n-type regions uncovered. On the surface of the silicon dioxide insulating layer, and between the n-type regions, an aluminium layer is deposited, which acts as the gate electrode. This method of insulating the gate is preferred to the use of a discrete wafer of insulating material, partly because of the thinness that can be achieved, and also because a thermally grown silicon dioxide layer passivates the silicon surface (i.e. it reduces very considerably the density of surface traps). Ohmic contacts are made to the n-regions (one of which acts as the drain and the other the source) and also to the gate. By making the gate positive with respect to the source, a positive bias exists between the gate and the p-type body in the region of the source. Positive charge carriers are repelled into the body and negative charge carriers are attracted to the surface. At the body/silicon-dioxide interface there is thus induced an n-type layer of mobile charge carriers. This layer connects the drain and source resistively; it is often referred to as an **inversion layer** because, on increasing the gate voltage from zero, the channel, originally p-type, becomes intrinsic and then finally an n-type layer is formed. Further increases in gate voltage increase the number of electrons in the channel, thus reducing the resistance between the source and drain. If a voltage is applied between the source and drain, a drain current, I_D, will flow. The magnitude of the drain current can be varied by applying varying voltages to the gate. Although the gate potential is positive relative to the source potential, no current is taken by the gate, since the silicon dioxide acts as an excellent insulator.

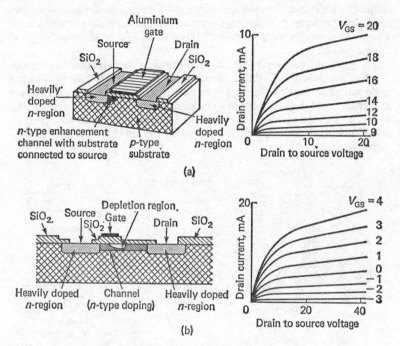

Fig. 35. The constructions of Hofstein and Heiman showing (a) an enhancement-type unit and (b) a depletion-type device.

Input resistances of the order of 10^{12} ohms have been achieved in available British units; whilst the Americans are claiming up to 10^{15} ohms for some of their transistors. There is thus available a solid-state device that is voltage-operated with a very large input impedance.

Insulated gate field-effect devices may be operated in one of two ways: namely, in the **enhancement mode** or in the **depletion mode**, depending upon the form of construction used. In the **enhancement** mode there is an *n*-type channel between heavily doped *n*-type regions with the gate extending across the entire channel as in Fig. 35(a). The gate is forward-biased, enhancing the number of electrons in the channel and reducing the source-to-drain resistance. At zero gate voltage the number of charge carriers in the channel is very low and so the drain current is effectively zero.

One of the disadvantages of the enhancement type unit is the large capacitance associated with the gate electrode. To overcome this an offset gate that does not cover the whole of the channel is used. Normally this would produce a very high resistance in the channel region not influenced by the gate. However, by suitable doping, a channel may be produced that has appreciable conductivity at zero gate voltage. Such a transistor is a **depletion** type and has the drain, source and channel regions all of the same conductive-type material although the drain and source regions are still heavily doped. The gate voltage must then be driven to some negative value before the drain current is zero. Fig. 35(b) shows the cross-section of this type of unit together with typical characteristics.

The pinch-off voltage, V_p, for a given transistor may be positive, zero or negative depending upon the construction. In practice, it is difficult to determine just when the drain current is zero so V_p is defined as that gate voltage that reduces the drain current to some specified low value (say 10 to 20 μA).

Fig. 36 shows an alternative geometry. Some manufacturers (e.g. Ferranti and Mullard) make a fourth connection to the substrate creating a four-terminal device.

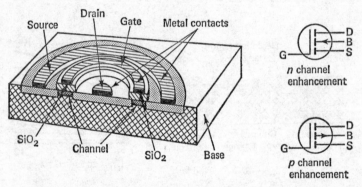

Fig. 36. Alternative geometry for an insulated gate f.e.t. The heavily doped regions are shown dotted.

Silicon Temperature Sensors

During investigations into silicon strain gauges it has been discovered that by a proper selection and cutting of a suitably prepared silicon crystal, a high-sensitivity solid-state temperature sensor could be produced. The resistance of such a sensor was found to vary in a linear way with temperature. The device is stable and has an extremely rapid response to changes of temperature (Fig. 37).

The silicon temperature sensor is a single crystal of almost pure silicon. Unlike the thermistor, with its compound oxide structure, the silicon unit can be manufactured with a high degree of reproducibility. The almost linear temperature/resistance graph is a distinct advantage over the logarithmic curve of the thermistor.

The sensors are available with an operating temperature range of about −170 to +300°C. They are normally calibrated over the range −10 to +170°C with an accuracy better than ±0·5°C. The theoretical response time is about five times faster than platinum for similar geometries. The relative

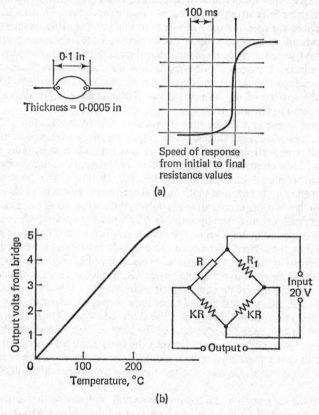

Fig. 37. Silicon temperature sensor: (a) dimensions together with graph showing speed of response; (b) calibration curve for sensor (R) when bridge connected. R_1 balances for zero output at the lower temperature (e.g. 0°C). For linearity K = 5.

resistance change over the range −10 to +250°C is from 40 to 50 per cent of the maximum resistance. The high sensitivity means that the output from the bridge circuit of Fig. 37(b) is high enough to drive milliammeters, measuring and control devices directly. An intermediate amplifier is often unnecessary.

The Semiconductor Gas Sensor

Although we have previously been discussing devices that depend upon single crystals of germanium, or more usually silicon, the reader should be aware that a large number of compounds are known which form crystalline semiconductors. Such compounds cannot at present yield satisfactory transistors, but other useful devices can be manufactured. It is a characteristic of semiconductors that their electrical conductivity depends upon such physical quantities as temperature, intensity of illumination, and magnetic field strength. The thermistor, already described, is a compound semiconductor whose resistance is strongly dependent upon temperature. Understandably, such devices are used for the measurement and control of temperature. The **magneto resistor** is another semiconductor device, made from a polycrystalline compound, which alters its resistance according to the strength of the magnetic field in which it is placed. Devices made from the compound cadmium sulphide change their resistance according to the intensity and type of light that falls on to the surface of the semiconductor. These **photocells** are discussed in greater detail in Chapter 14.

The **gas sensor** is a device that is probably new to most readers. Such a device can provide very valuable protection against fire and explosion in boats and yachts as well as buildings such as old people's homes. This device consists of two electrodes embedded in a bead of semiconductor material. The main constructional features are shown in Fig. 38. The electrodes are made from an alloy of iridium and platinum and each has a resistance of about 2 Ω. The bead is not a single crystal, but is polycrystalline (i.e. an agglomeration of many crystallites) and usually consists of an oxide of tin which has been suitably doped to make it an *n*-type material. When heated in air, oxygen is absorbed by the surface of the bead and a drastic reduction of the number of free electrons in the *n*-type material takes place. It is the function of one of the electrodes in the bead to act as the heater; when the bead comes into contact with a deoxidizing gas or vapour such as hydrogen, methane (natural gas), petrol fumes, coal gas and bottled gases such as propane and butane, absorption of the gas neutralizes, so to speak, the absorbed oxygen. The number of free electrons and hence the conductivity of the bead increases considerably. This change in conductivity is sensed between the two electrodes embedded in the bead. This device therefore acts as a gas-dependent resistor and can be used in circuits similar to those used for the thermistor—a temperature-dependent resistor.

The device is usually mounted on a base and protected with a flash-proof wire mesh or shield. This shield acts on the same principle as the miners' Davy lamp. The sensor works equally well on a.c. or d.c. and, if coupled to an amplifier, can be made the basis of alarm systems. The basic circuits are shown in Fig. 38, where R_L represents the circuit being operated. In the case of the d.c. circuit it will be seen that the current from the 0 V terminal, through the bead and R_L to the 12 V terminal, depends on the resistance of the bead. The output signal is developed across R_L. For domestic and marine applications it is usual to arrange that the associated circuitry trips the alarm at gas concentrations of only one tenth of the lower explosive level in the case of bottled gases and one twentieth of the explosive level for natural gas.

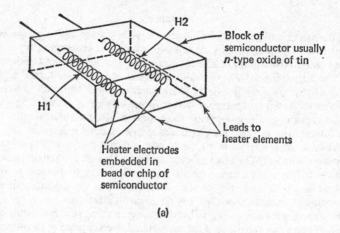

(a)

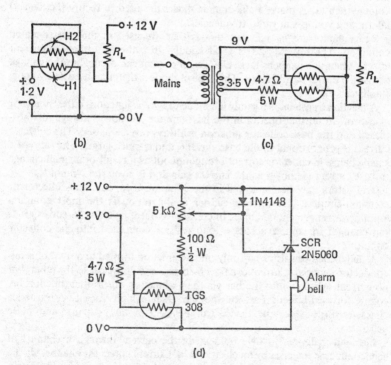

(b)

(c)

(d)

Fig. 38. The silicon gas sensor: (a) schematic diagram showing constructional features; (b) basic d.c. circuit; (c) basic a.c. circuit; (d) practical circuit for an alarm system to detect gas.

Summary

Transistors and other semiconductor devices are the **active components** of modern electronic apparatus. Atoms within a pure silicon crystal are arranged in a regular lattice and are bound together by strong covalent bonds formed by sharing electrons. A pure silicon crystal is a semiconductor with low conductivity. By doping with trivalent impurities, such as boron, the conductivity is increased and **p-type silicon** is formed. When pentavalent impurities such as arsenic or phosphorus are incorporated into the lattice of pure silicon, the conductivity is again increased and *n*-type material is formed.

When a single crystal is *n*-type at one end and *p*-type at the other a ***pn junction*** is formed. When this junction is **forward-biased** by applying a positive voltage to the *p*-type region and a negative voltage to the *n*-type region, current flows through the crystal. This current is considerable when the forward-bias exceeds about 600 mV in a silicon junction.

By applying an alternating voltage to the crystal, current flows only during the half cycle that the diode is forward-biased. **Rectification** of the alternating voltage then takes place. **Point-contact diodes** are used to rectify (i.e. detect) alternating voltages at radio frequencies.

Zener diodes are *pn* junction diodes that are used in the **reverse-biased** condition. The reverse-bias is great enough to cause an **avalanche** current effect. When this happens large changes of current can occur, but the voltage across the device hardly alters. This type of diode is therefore used as a voltage stabilizer.

A **junction transistor** is a sandwich made of two *pn* junctions either in *pnp* or *npn* form. In normal operation the base-emitter junction is always forward-biased and the base-collector junction is always reverse-biased. The collector current is proportional to the base current; and is much larger. The ratio of a small change in collector current brought about as a result of a small change in base current is known as the **current gain** and is given the symbol h_{fe}.

Transistors may be connected in the common-base, common-collector or common-emitter. The **common-emitter** configuration is the most common amplifier arrangement. Changes in collector current give rise to corresponding changes in voltage across a load resistor connected into the collector circuit.

A **unijunction transistor** has only one *pn* junction formed in a 'bar' of semiconductor material. Current can flow from one end of the bar to the other, but no current will flow into the bar via the *pn* junction unless that junction becomes forward-biased. Part of the bar then exhibits negative resistance characteristics; hence the device can be used as an oscillator and pulse generator.

Field-effect devices involve influencing the current through a channel of semiconductor material by an electric-field. Current enters the channel via the source lead, and leaves via the drain lead. The field is produced by applying a voltage to a third electrode called a **gate**. **Junction field-effect transistors** are made by forming a *pn* junction between the gate and channel. **Insulated-gate f.e.t.'s (M.O.S.T.'s)** have the gate electrically insulated from the channel.

Semiconductor devices that use polycrystalline material are silicon **temperature sensors**, whose resistance alters with temperature, and silicon **gas sensors**, whose resistance alters in the presence of hydrocarbon gases.

Questions

1. In spite of the fact that transistors are used in every form of modern electronic equipment, thermionic valves are still being used in the manufacture of television sets. Why should this be?

2. Explain what is meant by the terms '*p*-type' and '*n*-type' silicon. How does electrical conduction take place in these materials?

3. How does conduction take place in a crystal that contains both *p*- and *n*-type material?

4. Why is the common-emitter mode used much more frequently than the common-base mode in voltage amplifiers?

5. How can a 'hole' transfer charge within semiconductor material?

6. How would you set up an experiment in the laboratory to obtain the characteristics of a transistor?

7. Why does the resistance of copper rise with temperature whilst that of glass falls with temperature?

8. On circuit diagrams there are often arrows drawn to show the direction of the current. It is known that in wires the electron flow is in the opposite direction. Why is it not necessary now to change the arrow direction representing 'conventional' flow?

Suggested Further Reading

Olsen, G. H., *A Course Book for Students*, Butterworth, 1973.

Reddihough, J. A., and Sinclair, I. R., *Beginners Guide to Transistors*, Newnes–Butterworth, 1975.

4

AUDIO VOLTAGE AMPLIFIERS

Transducers are used in laboratories and industrial locations to convert the quantities under observation into suitable electrical signals. Rarely is sufficient output available to actuate an indicating device directly. The electric output from photocells, strain gauges, glass electrodes, thermistors, moving coil accelerometers, etc., all need to be increased, i.e. **amplified,** before the signal can be suitably recorded or indicated. Readers will be well aware of the fact that the output from a piezo-electric crystal pick-up cannot satisfactorily energize a loudspeaker directly. Some intermediate electronic apparatus must therefore be provided between the transducer and final indicator or load to effect the necessary amplification. The design of a suitable amplifier depends upon the transducer and the load, upon the required power output and on the frequency or nature of the signal.

Amplifiers may be classified according to the function they perform. If frequency is the classifying criterion then amplifiers are designated z.f., a.f. or l.f., r.f., v.h.f. or u.h.f. The z.f. **(zero frequency)** types are used to amplify steady voltages or currents. The term **d.c. amplifier** is often used for this type. Some ambiguity as to the meaning of d.c. arises, but most workers accept the term 'direct coupled'. We shall use the term 'd.c. amplifier' to mean an amplifier that can amplify steady voltages and signals whose frequency range extends down to zero frequency. All of the amplifiers for zero or very low frequency work are directly coupled between their stages.

A.F. (audio-frequency) amplifiers are used to amplify signals in the audible range, i.e. 20 Hz to 20 kHz (kilohertz). However, for faithful amplification of the signal waveform the amplifier must have a suitable response in a range of frequencies extending from about 10 Hz to 100 kHz. This is because non-sinusoidal periodic waves of audible frequency have harmonics extending beyond the upper limit of the audio range. The term **l.f. (low-frequency) amplifier** is to be preferred when audio work is not involved.

R.F. (radio-frequency), v.h.f. (very high frequency) and **u.h.f. (ultra-high frequency) amplifiers** are used when the signal frequency is much higher than the audio range. The r.f. signals extend over a range that includes the familiar long, medium and short-wave communication bands, say from 200 kHz to 30 MHz (megahertz, i.e. millions of cycles per second). The v.h.f. range, used for television and frequency modulated radio transmissions, goes up to about 200 MHz while the u.h.f. band—used for colour television, for example—extends over many hundreds of megahertz.

The dividing lines between the different bands are not clearly defined. The techniques used in the design and construction of amplifiers vary enormously depending upon the frequency range to be handled. In this chapter we shall concentrate exclusively on amplifiers for use in the audio range of frequencies. These amplifiers will incorporate discrete transistors and resistors (i.e. will use separate components) so that readers can appreciate the principles involved. In the next chapter, dealing with integrated circuits, we shall see how audio and other amplifiers can be constructed when most of the design work and construction is done for us. This means that instead of building amplifiers from the separate 'bricks', we build a system by buying and incorporating whole 'walls'.

An amplifier is a circuit that incorporates one or more transistors and is designed to increase an alternating signal applied to the input terminals. It is called a **voltage amplifier** if the size or magnitude of the output voltage is considerably greater than the input voltage. The ratio of the output voltage to the input voltage is called the **voltage gain** of the amplifier. The main function of a voltage amplifier is to produce a given gain with the minimum of distortion, i.e. the output voltage should have the same waveform as the input waveform, but should of course be much increased in magnitude. **Power amplifiers** are used to drive the output mechanism, e.g. a loudspeaker, a pair of earphones, a moving-coil meter or some other type of indicating device. The main function of a power amplifier is to deliver a good deal of undistorted power into the output device or load circuit; such an amplifier may often develop a voltage gain too, but this voltage amplification is of secondary importance.

When a number of amplifiers, each of which uses only one or two transistors, are connected in series (called a **cascade**) so that the output of one amplifier provides the input to the next amplifier, each of these amplifiers is called a **stage**. A complete amplifier consists of several stages; the earlier stages of an audio amplifier are designed as voltage amplifiers whilst the last stages form the power amplifier. (Power amplifiers are the subject of Chapter 6.)

The term 'audio' refers to sound as applied to human beings. The range of human hearing may be 20 Hz to 20 kHz in a young healthy ear, but as a person becomes older the upper limit falls. Most people over 50 years of age cannot detect sounds with a frequency content over about 12 or 13 kHz. It is unfortunate that because many young people today frequently 'listen' to the very loud sounds produced by high-power discotheques, they have hearing ability inferior to that of a healthy 60-year-old.

Electronic circuitry cannot of itself amplify the air pressure vibrations which we experience as sound. It is necessary to convert the pressure variations into electrical form by means of a **microphone**. Subsequently the electrical signals that represent the sound variations may be recorded on **discs** or **magnetic tapes**. To recover the information we use **pick-ups** to convert the groove modulations on discs into electrical form, or magnetic **heads**, as they are called, to convert the variations of magnetization on a tape into electrical form. The magnitude of the voltage produced by high-quality microphones is often only a few tens or hundreds of microvolts; high-quality pick-ups and tape-recorder heads rarely have outputs exceeding a few millivolts. Such tiny voltages must be amplified considerably before the programme signals can be fed into loudspeakers to produce realistic sound levels.

The Basic Voltage Amplifier

It will be recalled from the chapter on semiconductor devices that a transistor consists of two *pn* junctions fabricated within a single crystal in a three-layer configuration. Each layer is connected to the external circuit via leads, one of which is called the **emitter lead**, one the **base lead** and one the **collector lead**. If we forward-bias the base-emitter junction leaving the collector lead unconnected (or **open-circuit** as we say), then current can flow between the base and emitter leads. Alternatively, if we connect a battery between the collector and emitter leads leaving the base lead open circuit, we find that a negligible current flows between the two leads. This is because at least one of the *pn* junctions in the 'sandwich' must be reverse-biased. If we now connect a battery between the base and emitter leads so that the base-emitter junction is forward-biased, whilst simultaneously connecting a battery between the collector and emitter leads in such a way that the base-collector junction is reverse-biased, we can observe a remarkable result. The current injected into the base lead can influence a much greater current to flow in the emitter-collector circuit. The reasons for this have already been explained in the description of transistor action in Chapter 3. The remarkable result is that if the base current is increased by a certain proportion the collector current is also increased by the same proportion, so that changes in the base current give rise to corresponding, but much larger, changes in collector current; we have achieved **current amplification**.

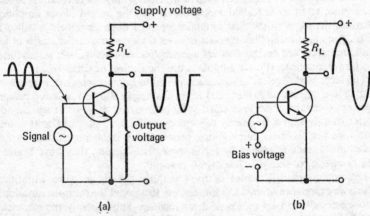

Fig. 39. (a) Transistor amplifier without bias. (b) Amplifier with bias. (a) shows the production of an output voltage that is severely distorted. This distortion can be largely avoided by applying a bias voltage as shown in (b). Using a bias battery is not practical so the circuits of Fig. 40 have been devised.

With the type of amplifier we are discussing here we are concerned not with current amplification, but with voltage amplification. We require to increase in size a small, varying or signal voltage, preserving as far as possible the waveform; we wish to avoid if possible any distortion of the waveform because in audio amplifiers distortion reduces the realism or fidelity of the reproduction of the programme material. We can convert changes in current into corresponding voltage changes by passing the current through a resistor. Since by Ohm's law $V = IR$, then for a fixed value of R any changes in I, the current, will bring about corresponding changes in V, the voltage. To achieve this in transistor amplifiers we connect a resistor between the collector lead and the battery supply, as shown in Fig. 39. This resistor is known as the **load resistor.** Any changes in current through the resistor produce changes of voltage across the resistor. Since the voltage at the positive supply lead is fixed, the changes in voltages must, and do, occur at the end connected to the collector. The collector lead is therefore the **output terminal** of the amplifier.

Biasing Circuits

We find that if we produce changes of collector current by connecting the signal source directly between the base and emitter of the transistor, then severe distortion is observed in the output voltage waveform. This is because the signal source cannot maintain the base-emitter junction in a forward-biased condition. If the signal amplitude is large enough there will be forward-bias during the time that the signal voltage is positive-going, but a reverse-bias condition is set up when the signal voltage is negative. The result is that the transistor is operating only during the positive portions of the signal voltage, and is cut off for the remaining time. An output voltage will therefore be produced for only the positive portions of the signal; for the remaining time the output voltage will be zero.

All voltages are measured with respect to the earth line, which in this case is connected directly to the emitter. The emitter lead is seen to be common to both the input circuit and the output circuit, hence the transistor is being operated in the common emitter mode. This mode is by far and away the most usual to use for small-signal voltage amplifiers.

The way out of the difficulty associated with the severe distortion is to arrange that even when the signal voltage is zero or less, the voltage between the base and emitter is such as to keep the base-emitter junction forward-biased. Such a voltage could be provided by a battery as shown in Fig. 39(b). The signal voltage is then superimposed on the battery voltage. The battery voltage is referred to as a **bias voltage** because the signal voltage, instead of varying about zero volts, is shifted or biased so that the variations are about some positive voltage. (For *pnp* transistors the bias voltage would be negative.) It is not very practical to use an actual battery to produce the bias voltage. Apart from the bulk and cost of the battery it will be seen that neither terminal of the signal source is connected to earth. This can cause difficulties especially when the signal source is a transducer such as a microphone or pick-up.

Fig. 40 shows the way in which suitable biasing voltages can be obtained. In Fig. 40(a) the base of the transistor is connected to the positive supply line via a high-value resistor; the signal source is isolated from the bias circuit by a capacitor. Any steady component in the signal is thus blocked out by the capacitor; only the useful signal variations (i.e. the a.c. component) can be transferred. It will be seen that even when the signal voltage is zero a steady

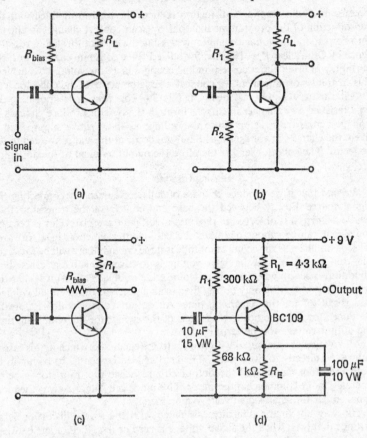

Fig. 40. Biasing circuits for a common-emitter transistor amplifier stage.

stream of holes can be injected into the base via R_{bias}. (Actually, electrons will be the only charge carriers travelling along the lead wires and through the resistor to the positive supply line; but for every electron that leaves the base and enters the base lead, an equivalent positive charge will be created in the base near to the connecting lead. This is equivalent to injecting a hole into the base.) We are thus forward-biasing the base-emitter junction by connecting the p-type base, via R_{bias}, to the positive supply, and the n-type emitter to the negative supply line. Another method is shown in Fig. 40(b). The base is connected to a potential divider and the correct values of R_1 and R_2 are chosen so that the biasing voltage is correct.

In practice the feedback arrangements shown in Figs. 40(c) and (d) are used. These arrangements are used for two reasons. First, the collector current will rise if the temperature of the transistor increases. Thus we have the wanted collector current (due to the biasing and signal currents being amplified by normal transistor amplification) together with the unwanted current which is due to the temperature rise. If the unwanted component is large compared with the wanted component not only is the performance of the amplifier degraded, but in addition the increased collector current heats up the transistor, thus causing an even greater unwanted current to flow. In severe cases, thermal runaway occurs and the transistor is destroyed. Although this effect can be severe in germanium transistors, silicon devices are much less affected by temperature rises. Even so, we still employ the more complicated bias circuitry. This brings us to the second reason for doing so. It may well be that for some reason the transistor has to be replaced. Transistors of the same type cannot be expected to have closely similar characteristics because of the difficulties inherent in their manufacture. Consequently, the replacement transistor may have a much greater unwanted current component than the device being replaced. To avoid the difficulties that this would cause we use the more complicated bias circuits even with silicon transistors.

Both of the circuits shown in Figs. 40(c) and (d) use a feedback stabilizing arrangement. In Fig. 40(c) this is achieved by connecting the bias resistor to the collector lead instead of the positive supply line. The bias current is determined by the resistance of the bias resistor and the voltage across it. Once the resistance has been determined the bias current is determined by the voltage on the collector. Suppose, after having designed and assembled the circuit, we need for some reason to change the transistor and the new one gives a higher standing current than the original one. The greater standing current (often called the quiescent current) causes a greater voltage drop across the load resistor in the collector line. Consequently the collector voltage falls; the voltage across the bias resistor, R_{bias}, also falls, as does the bias current. Since less current is now entering the base, the collector current falls. The fall in collector current largely offsets the original rise and so the collector current is found to be close to its original value even though we have changed the transistor. The mechanism described above also operates if the rise in collector current is due to increased operating temperatures. Falls in collector current are compensated for in a similar way.

The design of Fig. 40(c) is quite simple. Suppose we wished to design a single-stage common emitter audio amplifier of the type shown, then we may use any general-purpose silicon transistor. A 2N3704 or a BC109 may be used, for example. These transistors operate satisfactorily with a quiescent current of 1 mA. ('Quiescent' means quiet, so in this context the term 'quiescent' means the current present in the absence of an audio signal.) The least distortion will be produced if the collector voltage is half the supply voltage. If the supply voltage is, say, 9 V then the collector voltage should be 4·5 V. When the collector voltage is made to vary by applying a signal to the base, this voltage can vary equally about the 4·5 V value to a higher or lower voltage. We see that if a quiescent collector voltage of, say, 8 V was chosen then it would only be possible to rise by 1 V before 'bumping' the positive supply voltage. Having fixed a quiescent collector voltage of 4·5 V, we see that 4·5 V is dropped across the load resistor R_L. Since the quiescent current is

1 mA, then $R_L = 4.5$ V/1 mA $= 4.5$ kΩ. Now 4·5 kΩ is not a preferred value so we have to choose the nearest preferred value, namely 4·7 kΩ.

The manufacturer's data says that a BC109 has a typical current gain, h_{fe}, of 400. A reasonable approximation to the bias current is therefore the quiescent current divided by the current gain, i.e. 1 mA/400 $= 2.5$ μA. The base-emitter junction must be forward-biased by 650 mV (i.e. 0·65 V) for correct operation, so we see that the voltage across R_{bias} must be $4.5 - 0.65 = 3.85$ V. The current through R_{bias} is to be 2·5 μA, so the bias resistor must have a value of 3·85 V ÷ 2·5 μA $= 1.54$ MΩ. We would therefore choose the nearest preferred value of 1·5 MΩ.

In order not to disturb the bias current by any steady component in the signal (and bearing in mind that only the variations in the signal are to be amplified) any steady component in the signal is blocked out by a capacitor. Only the varying component is therefore transmitted. The size of the capacitance used is determined by the input time constant, which is the product of C and the input resistance. We find that a small-signal transistor of the BC109 type has an input resistance of about 2 kΩ. It will be recalled from the work we did in Chapter 2 that satisfactory transmission of a signal waveform would occur only if the coupling circuit has a time constant of 5 or 10 times the period of the signal. Let us say that the lowest frequency we wish to handle is 40 Hz. The period of such a signal is 1/40 seconds. Hence our time constant should be, say, 5 times this, namely 1/8 second, i.e. $CR = 0.125$. If R is 2,000 then $C = 0.125/2,000 = 62.5$ μF. The capacitance is not critical and a value of 50 μF would be perfectly satisfactory. Since the working voltage is low the electrolytic capacitor would be physically quite small.

Graphical Techniques

Although this book is intended as a 'made simple' text there may be some readers, especially students, who would like to know about graphical techniques and what are called **load lines**. Those not interested can skip this section without affecting their introductory understanding of amplifiers.

When we plot a graph showing how the collector current varies with the collector-emitter voltage for any fixed value of base current, the form of the graph is that shown in Fig. 41. Here several graphs of this type are drawn on one set of axes, each graph being for a different base current. The family of curves is known as a **set of output characteristics**. For any value of base current and collector voltage we could read off the corresponding value of collector current. Consider now the circuit diagram accompanying the set of output characteristics. Here we have a fixed supply voltage, V_{CC}, and a load resistor, R_L, in the collector circuit. For any particular value of collector current a voltage drop across R_L will be experienced which is equal to $I_c R_L$. The sum of this voltage and the collector voltage, V_{ce} is equal to the fixed supply voltage V_{CC}. Hence

$$V_{CC} = V_{ce} + I_c R_L$$

Now the characteristic curves are plotted on axes which show I_c as the y-axis and V_{ce} as the x-axis. We can transpose the above equation, using simple algebra, so that we have I_c on one side and everything else on the other side of the equation. This equation then becomes

$$I_c = -\frac{1}{R_L} V_{ce} + \frac{V_{CC}}{R_L}$$

which is in the form $y = mx + c$, where y is I_c, m (the slope) is $-1/R_L$ and the intercept on the y-axis is V_{CC}/R_L. This straight-line graph can be super-imposed on the set of output characteristics as shown; the straight line is known as the **load line**. From it we can make a more accurate estimate of the bias current needed as well as making many other deductions which we cannot go into in this book. (See Suggested Further Reading.)

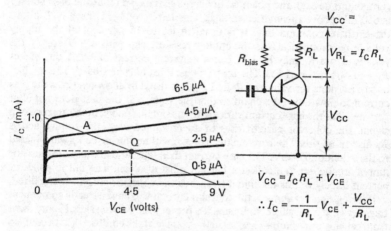

Fig. 41. The simplest transistor amplifier.

The estimate of the bias current is made in the following way. We must choose an operating point, Q, which is in the middle of the load line. This will enable the collector current to vary, and hence the collector voltage to vary, when the signal source changes the base current. If the operating point were too near the I_c or V_{ce} axes then as the base current varies with the signal we then may get down to zero collector current or almost zero collector volt-age; it would obviously not be possible to drive the transistor with less than zero collector current. Severe distortion of the waveform would result if we did. Hence the operating point must be kept at or near the centre of the load line. In the absence of any signal we will be at the quiescent operating point, Q.

In our example we see that the corresponding collector voltage is 4·5 V and the collector current is 1 mA. The load line intersects the 2·5 μA base current at the point Q: 2·5 μA is therefore the bias current that must flow when the signal is absent. When the signal causes the base current to rise, say, to 4·5 μA the collector current rises and so the collector voltage falls because of the increased voltage drop across R_L. We now arrive at the point A. When the base current is driven by the signal to less than the bias current the collector current falls and consequently the collector voltage rises. Thus we see that when the signal voltage rises and falls, causing the base current to rise and fall, the collector current falls and rises. The collector, or output, voltage is therefore 180° out of phase with the signal voltage. We see also that we can predict the output voltage swing if we know the input current swing.

Fig. 40(d)—see page 76—shows a different, and often used, method of obtaining the correct biasing conditions. The base is biased via a potential divider network R_1, R_2 and stabilization against thermal and other variations is achieved by including an emitter resistor R_E in the emitter lead. The stabilizing action is as follows. Remembering that we must forward-bias the base-emitter junction by 650 mV for a silicon transistor (150–200 mV for a germanium device), the potential divider is designed so that the base voltage is 650 mV above the emitter voltage. This latter voltage is itself $R_E I_c$ above the earth or common line. It is usual to let 10–15 per cent of the supply voltage be dropped across the emitter resistor, e.g. with a 9 V supply the voltage across R_E may be 1 V. For a quiescent current of 1 mA this makes $R_E = 1$ V/1 mA $= 1$ kΩ. The base voltage must now be 1·65 V. This enables us to determine the values of R_1 and R_2. It is usual to allow ten times the bias current to flow down the chain so that the voltage at the base does not alter very much as the base current varies with the signal. If now, in the absence of a signal, the collector current rises in an unwanted way, the voltage across R_E also must rise. The base-emitter voltage will no longer be forward-biased to the extent of 650 mV. The base-current then falls, and because of transistor amplifying action, the collector current must also fall. The fall in collector current largely offsets the initial rise; thus the collector current is stabilized.

If now a signal is applied, the collector current will have a steady component together with a varying component. To prevent the varying signal component from causing undesirable changes in the voltage at the emitter, R_E is decoupled by placing a capacitor with a large capacitance in parallel across it. The a.c. variations are therefore shorted out, so only the steady component is involved in the biasing mechanism.

Multistage Amplifiers

When more gain is required than can be obtained from a single stage of amplification, then two or more stages can be coupled. Fig. 42 shows how two stages can be connected together. Direct connection between the collector of the first transistor and the base of the second transistor is not possible without upsetting the bias conditions in the second stage. This is because the steady voltage at the collector is much higher than that required at the base of the second transistor for correct biasing. The difficulty can be overcome by interposing a coupling capacitor C_c which blocks off the steady voltage, but allows signal voltages to be transferred.

At first sight it may appear that the overall gain of the two-stage amplifier is found by multiplying the gain of the first stage by the gain of the second stage. This is the correct procedure provided it is realized that the gain of the first stage when connected to the second stage is lower than it would be if the first stage were operating in isolation. In isolation the load of the first stage (which determines the gain) is the collector resistor. However, when two stages are coupled, the first stage transistor has to drive not only the collector load resistor, but also the second stage. The effective gain of the first stage is therefore reduced.

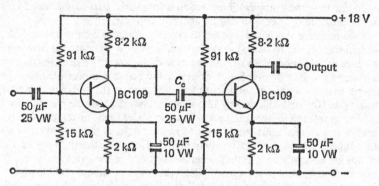

Fig. 42. Two-stage transistor amplifier.

Frequency Response and the Decibel Scale

For any amplifier there is a range of frequencies over which the gain of the amplifier is constant. This range is called the **mid-frequency range**. All amplifiers are designed to operate satisfactorily over this range of frequencies. However, above and below the mid-frequency range the gain becomes smaller. At the low-frequency end of the spectrum the fall in gain is due mainly to the coupling capacitor. As the frequency becomes lower and lower the reactance of the capacitor becomes larger and larger. Eventually it is so large that very little of the low-frequency part of the signal is transferred. At high frequencies the stray capacitances form leakage paths. These stray capacitances are formed between the signal lines and the metal chassis or housing case, and also between the signal lines and other connectors, say, on the printed circuit board. Although such capacitances are small (perhaps only 20 or 30 pF) and are of no consequence at low frequencies, as the frequency of operation increases the strays become more and more important. For example, the reactance $[1/(2\pi fC)]$ of 30 pF at 1 kHz is about 5·3 MΩ, but at 10 MHz it is only about 530 ohms. The signal therefore 'leaks away' to the earthy parts of the circuit and the gain at the higher frequencies falls.

The performance of an amplifier is often illustrated graphically by what is called a **frequency-response curve**. This curve is a graph showing how the response (plotted on the y-axis) varies with frequency (plotted on the x-axis).

Since it is very useful to compare the frequency responses of various amplifiers, it is not so important to know the actual gains as it is to know the comparative or relative gains. To enable us to do this it is universal to use what is called the **decibel scale**. The bel scale was introduced originally by telephone engineers to express the ratio of two powers, one at each of two points in the telephone line. The scale was named in honour of Alexander Graham Bell, the pioneer in telephone work. The bel is in practice too large a unit for electronic and acoustic purposes and so the unit **decibel** (dB) was introduced, one decibel being one tenth of a bel. Because such an enormous range of the ratios of powers is encountered we use a logarithmic scale. In this way we avoid crushing the graphical information into a small part of the scale. This can be illustrated by considering an acoustic example.

If we measure the power needed to create a sound which is just audible, we find such power to be very small indeed. If we now compare the powers to produce other sounds with the minimum audible power, we discover that an enormous range is involved. For example, the acoustic power near to a jet engine may be a million million times greater than the minimum audible power (i.e. 10^{12}, or 1 followed by 12 noughts). Consider now the position if we tried to graph the power using an arithmetic scale. Eight or ten centimetres (say) of the vertical scale would have to represent many millions to cover all the sound levels that could be encountered. However, the majority of the sounds we normally hear in the home, at work and in the street are normally only a few thousand (not many millions) times the minimum audible power. To plot these everyday sounds a small band not much thicker than a pencil line would cover the normally encountered sounds. Clearly we need to 'stretch out' this part of the scale, whilst compressing that part of the scale that deals with the loud sounds. A decibel scale does just that.

For electronic amplifiers we find the decibel scale to be just as useful. Here we are comparing different output voltages at different frequencies, whereas in acoustic work we compare different sound pressure levels. When comparing the ratios of two voltages we find that here also an enormous range is encountered. Hence instead of plotting these ratios on an arithmetic scale we use the decibel scale.

To express the ratio of two voltages in decibels we take the logarithm to the base 10 of the ratio and multiply this logarithm by 20. (The reason we use 20 is that we are comparing voltages, not powers. Since power is proportional to the square of the voltage we should take the logarithm of the square of the two voltages and multiply by 10. However, when we wish to obtain the logarithm of the square of the ratio of the two voltages we look up the log. of the ratio and double it. Hence we must double the 10, which becomes 20.) So the response (R) in decibels is found by using the formula

$$R(\text{dB}) = 20 \log_{10} \frac{V_f}{V_s}$$

where V_f is the output voltage of the amplifier at some frequency and V_s is the output voltage at a standard frequency—usually 1 kHz for audio work.

Negative-Feedback Amplifiers

Ordinary amplifiers of the type already discussed suffer from various forms of distortion, and their performance is altered by the ageing of components and variations of supply voltages. Straightforward amplifiers are not therefore suitable either as measuring amplifiers or for audio work. All modern high-performance amplifiers employ the principle of **negative feedback**, a system whereby a fraction of the output voltage is fed back to the input in antiphase.

We have seen previously that with a single-stage amplifier a rise in base voltage produces a rise in collector current, which in turn produces a fall in collector voltage. The collector voltage, which is the output voltage, is out of phase by 180° since when the input voltage is rising the output voltage is falling; this is what we mean by **antiphase**. An amplifier must therefore have one, three or an odd number of stages to produce a 180° phase shift between the input and output voltages.

Distortion

The output of an amplifier is said to be **distorted** if a change of waveform occurs between the input and output terminals. (This is not the same as a change of phase. The input and output voltages may each have the same general shape, or waveform—in which case there is no distortion—but the waveform of the output voltage may be shifted relative to the input voltage in a time sense.) Distortion arises when the output waveform contains frequency components not present in the original signal, or, where complex signals are involved, when the phase relationship between the various components of the signal may be altered. The relative amplitudes of these harmonic components may also be altered. We will now consider the reasons why simple amplifiers introduce distortion.

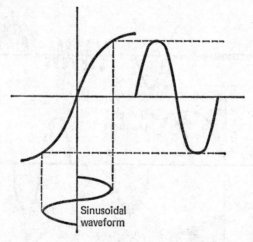

Fig. 43. Curvature of the transfer characteristic leads to the introduction of harmonic distortion.

The actual output of an amplifier is necessarily limited. Although the gain of an amplifier may be, for example, 1,000, this does not imply that any magnitude of input voltage is amplified 1,000 times. An input voltage of 100 V (r.m.s.) does not produce an output voltage of 100 kV in the types of amplifier we are discussing. There is a linear relationship between output and input voltages only over a restricted operating range where overloading of any stage in the amplifier is absent. The relationship between input and output voltages is known as the **transfer characteristic** (Fig. 43), and this characteristic is curved at the ends. The gain of the amplifier therefore varies with the instantaneous magnitude of the input signal, and **non-linear distortion** is said to be present. Curvature of the dynamic characteristics of the transistors contributes to non-linearity of the transfer characteristics. The application of a sinusoidal input voltage results in a periodic output waveform that is non-sinusoidal. Fourier analysis shows that spurious harmonics are present, the result being known as **harmonic distortion.** Total harmonic distortion, D, is measured as the root of the sum of the squares of the r.m.s. voltages of the individual harmonics, divided by the r.m.s. value of the total signal, V, i.e.

$$D = \frac{\sqrt{(V^2_{H2} + V^2_{H3} + V^2_{H4} \ldots V^2_{HN})}}{V} \times 100 \text{ per cent}$$

where V_{H2}, V_{H3}, etc., are the r.m.s. values of the harmonic components.

Intermodulation distortion (Fig. 44) is a form of non-linear distortion whereby the amplification of a signal of one frequency is affected by the amplitude of a simultaneously applied signal of lower frequency. Combination frequencies are produced which have values equal to the sum and difference of the two applied frequencies.

Attenuation distortion is caused by the variation of the gain of an amplifier

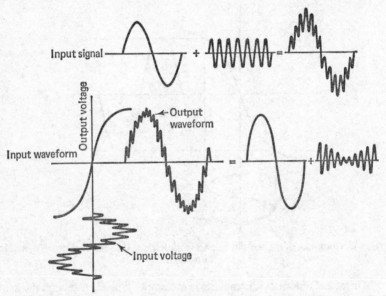

Fig. 44. When two sinusoidal input voltages are applied simultaneously, the non-linear transfer characteristic reduces the amplitude of the higher-frequency signal at times when the lower-frequency signal is near to the maximum and minimum voltages. This is intermodulation distortion.

with frequency. If, for example, a complex waveform has a harmonic of a high frequency and the gain at that frequency is very low, clearly this harmonic must be almost absent in the output waveform.

Phase distortion is present when the relative phases of the harmonic components of a signal are not maintained. Such distortion is caused by the presence of reactive and resistive components in the circuit. In cathode-ray oscilloscope amplifiers, television video amplifiers and in radar circuits, phase distortion is highly undesirable. It is often said that phase distortion is unimportant in audio amplifiers as the ear is insensitive to moderate changes in phase. Whilst it is true that the ear is insensitive in this respect, it is not true that demands on the audio amplifier can be relaxed. The effect of phase shift is of great importance to the speaker diaphragm from the transient point of view. The quality of a sound depends, among other things, upon the attack and decay times. To obtain similar attack and decay times in the reproduced sound, phase distortion should be reduced to a minimum.

Transistor and circuit noise, and 50 Hz components ('hum') are usually classified as distortion when introduced by an amplifier into a signal otherwise free of them.

Principle of Negative Feedback

Most forms of distortion may be markedly reduced by using negative feedback. **Feedback** is said to occur in amplifiers when part of the output voltage of the amplifier is added to, or subtracted from, the input signal. When a fraction of the output is added to the input signal the feedback is said to be **positive.** The gain of the amplifier rises usually in an uncontrolled way and oscillations occur. This is the subject of a later chapter.

When part of the output is fed back to the input in antiphase (i.e. 180° out of phase), then subtraction occurs and the feedback is said to be **negative.** Fig. 45(a) is one way of representing a feedback amplifier when we wish to make a quantitative examination.

The main amplifier (of the straightforward type previously discussed) is represented by the triangle and has a gain of A. A fraction of the output voltage, β, is selected by a suitable circuit and fed back to the input. So far as the main amplifier is concerned, it 'sees' an input voltage, V_g, which is the signal voltage V_s plus the voltage feedback, βV_{out}, where V_{out} is the output voltage. Therefore

$$V_{out} = AV_g = A(V_s + \beta V_{out})$$

$$V_{out}(1 - \beta A) = AV_s$$

$$\therefore V_{out} = \frac{A}{1 - \beta A} V_s$$

This is the general feedback equation. The usual way of obtaining negative feedback is to arrange for a 180° phase shift in the amplifier, i.e. A is negative. The equation is then

$$V_{out} = \frac{-A}{1 + \beta A} V_s$$

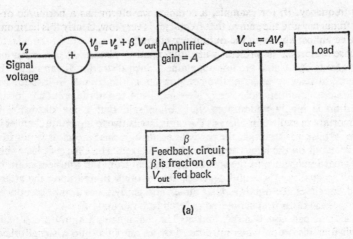

(a)

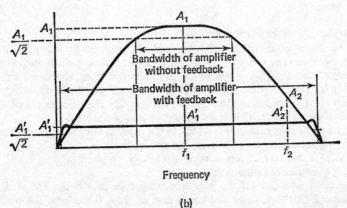

Frequency

(b)

Fig. 45. (a) Block diagram representation of a feedback amplifier. (b) The effect of negative feedback on bandwidth. Note the effect on gain at different frequencies. Without feedback the amplifier has gains of A_1 and A_2 at f_1 and f_2 respectively. A_1 is much greater than A_2. With feedback, however, the gains at f_1 and f_2 are A'_1 and A'_2. These gains are equal.

i.e. the gain with negative feedback, A', is given by

$$A' = \frac{-A}{1 + \beta A}$$

From this it can be seen that the gain of the feedback amplifier is lower than that of the main amplifier by a factor $1 + \beta A$. This is the price that must be paid to obtain the advantages described below. The product βA is the gain around the feedback loop and is, therefore, called the **loop gain**.

It is easy to arrange that the gain of the main amplifier (called the **open loop gain**) is very high (say 10^5 or 10^6). The loop gain is much greater than unity with the values of β used in practice. The effective gain of the negative feedback amplifier therefore closely approaches $-1/\beta$, i.e.

$$A' = \frac{-A}{A\beta} = -\frac{1}{\beta}$$

The gain of the feedback amplifier (called the **closed loop gain**) is thus independent of the gain of the main amplifier provided the latter is large. Variations of supply voltages, ageing of components, and other causes of the variations in gain of the main amplifier are therefore relatively unimportant in a negative-feedback amplifier. The gain with feedback depends only on β and this can be made very stable by choosing simple feedback circuits that employ very stable circuit components.*

Effect of Negative Feedback on Gain Stability

Let us suppose that the gain of an amplifier is -10^6 and that $\frac{1}{100}$th of the output voltage is fed back in antiphase, i.e. $\beta = 10^{-2}$. The gain of the negative feedback amplifier is then

$$A' = \frac{-10^6}{1 + 10^6 \times 10^{-2}} = -100 \text{ (near enough)}$$

If now a serious upset in the main amplifier reduces the gain from 10^6 to 10^4, the gain of the negative feedback amplifier becomes

$$A' = \frac{-10^4}{1 + 10^4 \times 10^{-2}} = -100$$

which is the same as before. The gain of the feedback amplifier has not been altered by a large change in the gain of the main amplifier. This independence of gain results from the fact that the input to the main amplifier is the difference between the signal voltage and the voltage fed back. If the gain in the main amplifier falls, the difference voltage will increase slightly and so the output remains almost constant.

Effect of Negative Feedback on the Frequency Response

The upper curve of Fig. 45(b) represents the frequency response of a straightforward amplifier. It has a gain of A_1 at frequency f_1 and a gain of A_2 at f_2. A_2/A_1 is small, resulting in a restricted **bandwidth**. (The bandwidth of an amplifier is the range of frequencies over which the gain of the amplifier does not fall more than 3 dB below the maximum gain.) If now negative feedback is applied, the gain at f_1 is $A'_1 = -A_1/(1 + \beta A_1)$ and at f_2 is $A'_2 = -A_2(1 + \beta A_2)$. Therefore

$$\frac{A'_2}{A'_1} = \frac{-A_2}{(1 + \beta A_2)} \frac{(1 + \beta A_1)}{-A_1}$$

---

* The use of β is conventional and was in common use before transistors were invented. The significance of the symbol should not be confused with that used for current gain in a transistor.

When βA_1 is much greater than 1 and βA_2 is also much greater than 1, then $A'_2/A'_1 \approx 1$. In other words, the gains at the two frequencies are approximately equal. This is shown in the lower curve of Fig. 45(b).

Negative feedback thus increases the bandwidth of the amplifier. One simple practical way of doing this over a single stage is merely to omit the emitter bypass capacitor. As the signal voltage rises the collector current rises. There is an increase in the voltage across the bias resistor resulting in the application of a smaller bias voltage. This offsets to some extent the rise in the signal voltage, and the base-to-emitter voltage is not then as great as it otherwise would have been. As the feedback is effective over a very wide frequency range, the benefits of negative feedback are obtained over entire operating frequency range. There is thus an increase in the bandwidth, greater gain stabilization and a reduction of distortion when compared with an amplifier that does not use negative feedback.

Input and Output Impedances

Although we need to know the voltage gain and distortion levels in an amplifier, there are other amplifier properties that often need to be considered. Of these, two important ones are the **input impedance** and the **output impedance**.

The **input impedance** is simply defined as the ratio of input voltage to input current. A knowledge of the size of the input impedance is needed since the amplifier must be driven by a signal source of some kind. Some signal sources such as crystal microphones and photodiodes can deliver only tiny currents because of their high internal impedance. Any attempt to amplify the signals from such sources will be successful only if the input impedance of the amplifier is high. This can be seen from Fig. 46. If the crystal microphone has

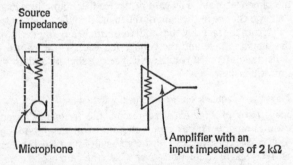

Fig. 46. Matching an amplifier to the transducer.

an internal impedance of 1 MΩ and the amplifier an input impedance of 2 kΩ (which is typical of the single-stage amplifier of the type we have already discussed), then, of the voltage generated by the microphone, only 2 kΩ/1 M, i.e. $\frac{1}{500}$, of this voltage is all that is available at the amplifier input terminals.

In order to have a reasonably large percentage of the developed voltage available at the amplifier input terminals the input impedance must be about 1 MΩ at least. One way of increasing the input impedance is to connect two

transistors in what is called a **Darlington pair** arrangement, shown in Fig.
47(a). Here the collector current of transistor T_1 is approximately h_{te} times the
input current. This collector current forms the base current of T_2. The collector current of T_2 is approximately the current gain (h_{te}) of T_2 times the base

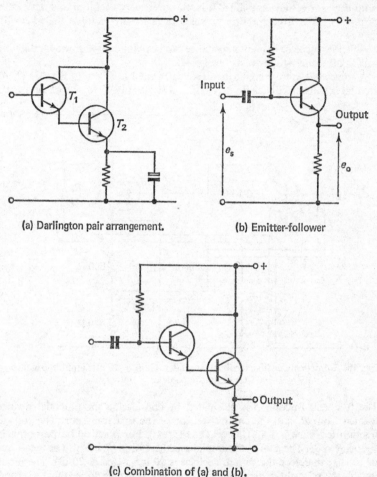

(a) Darlington pair arrangement.

(b) Emitter-follower

(c) Combination of (a) and (b).

Fig. 47. Methods of obtaining high input impedances.

current into T_2. The overall current gain is therefore $(h_{te})^2$ when each transistor
has the same current gain. The result is that only a tiny current is taken from
the source; in other words the input impedance is very high.

The input impedance can be increased by using negative feedback. A simple
feedback arrangement known as the **emitter-follower** is shown in Fig. 47(b).
In this arrangement the load resistor is transferred from the collector circuit

to the emitter circuit, and the output is taken from the emitter. The input voltage to the transistor is then the signal voltage, e_s, minus the output voltage, e_o. Thus we have 100 per cent negative feedback, i.e. $\beta = 1$. The voltage gain of this stage is therefore 1 (or slightly less in practice), so the arrangement is no use as a voltage amplifier. It is, however, very useful in that very little current is taken from the driver, which means that the input impedance is very high.

A further increase in input impedance can be obtained by using a Darlington pair as an emitter-follower as in Fig. 47(c).

A somewhat more complicated feedback circuit is shown in Fig. 48. There are two negative feedback loops. Loop 1 gives negative feedback of the signal.

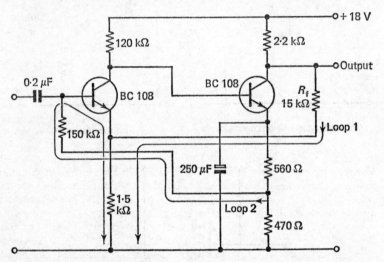

Fig. 48. A practical negative feedback amplifier. Gain = 20 dB; input impedance = 140 kΩ; output impedance = 140 Ω.

The feedback fraction β is calculated by considering the potential divider formed from R_f and the emitter-resistor of the first transistor. The voltage fraction fed back is $1{\cdot}5$ kΩ$/(1{\cdot}5$ kΩ $+ 15$ kΩ$)$. For practical purposes this is $\frac{1}{10}$, i.e. $\beta = \frac{1}{10}$. The gain of the amplifier is therefore $1/\beta$, i.e. 10. The logarithm of 10 is 1; therefore the gain in decibels is $20 \log_{10} 10$, i.e. 20 dB. The input impedance of such an arrangement is 140 kΩ. It will be noticed that a second feedback loop exists. This feedback does not operate on the signal, however, because of the presence of the 250 μF emitter-capacitor, which gives a signal or a.c. short to ground of the emitter. The purpose of this second loop is to stabilize the d.c. conditions.

The impedances of output devices are frequently quite low. In the case of loudspeakers, for example, impedances of 4 or 8 ohms are common. When the output impedance of the generator that is driving a load is fixed, it can be shown that maximum power is supplied to the load when the impedance of the load is equal to the generator's internal impedance. To obtain maximum power transfer we match the load and generator impedances by varying the load impedance, if that is possible. If the choice of load impedance is not available a transformer can often be used to effect the matching.

In practice it is often the case that the load impedance is not under our control. We may, for example, have to drive a given servomotor or a specified loudspeaker. Fortunately, when the power source is an electronic amplifier, the output impedance can be varied by using circuit techniques that involve feedback. When the generator impedance is variable and the load impedance is fixed then maximum power is delivered to the load when the generator impedance is zero. In audio work we find that the performance of a loudspeaker is improved if the output impedance of the amplifier is very low compared with that of the loudspeaker.

Frequency Selective Feedback

The types of negative feedback so far discussed have not been frequency selective because the feedback network has consisted of only resistors. The impedance of a resistor is the same at all frequencies and is merely the resistance of the component. When capacitors are used in the feedback path we find that the amount of negative feedback is not the same at all frequencies because the reactance of a capacitor varies with frequency. The combination of capacitors and resistors can form feedback paths and provide us with a means of controlling the frequency response of an amplifier (see Chapter 13). We shall content ourselves here with observing that for reproduction from both discs and magnetic tapes it is necessary to use an amplifier that amplifies the low frequencies more than the higher frequencies. This object can be achieved by using a **frequency-selective negative feedback** network. Two such arrangements are shown in Fig. 49 and 50.

To understand how they work we must remember two things: first that the opposition to alternating current flow (the reactance) of a capacitor increases as the frequency of the current decreases, and second that the gain of a negative feedback amplifier decreases as we increase the amount of feedback.

The necessary boosting of the bass frequencies can be obtained by modifying the feedback path of Fig. 48, replacing the resistor R_f with the capacitor/resistor network as shown in Fig. 49. At high frequencies the reactance of the capacitors is low and hence there is a good deal of feedback. This results in low gain at high frequencies. As the frequency of operation is reduced to the low audio frequencies the negative feedback is reduced because of the rise in the reactance of the capacitors. There is thus a consequent rise in gain at low frequencies. Circuit component values are carefully chosen so that the correct amount of boosting is obtained at specific frequencies. Modifications of the circuit values shown in Fig. 48 are also necessary to make sure that the input impedance of the amplifier is satisfactory for use with magnetic pick-ups.

Fig. 50 shows a feedback pre-amplifier for use with tape recorder heads. Boosting of the low frequencies is still required, but the shape of the required response curve is different, hence the different feedback network.

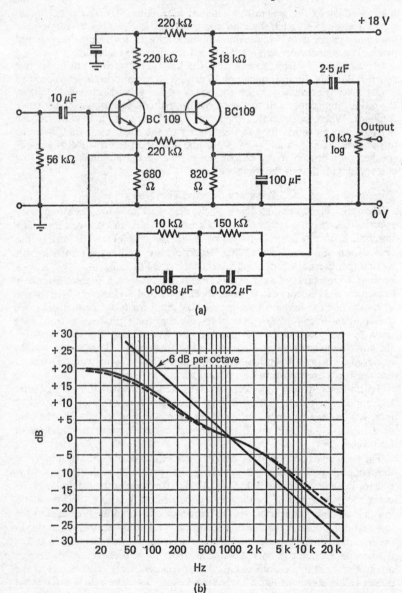

(a)

(b)

Fig. 49. A preamplifier for use with magnetic pick-ups. (a) A preamplifier that boosts the bass frequencies in a way suitable for use with magnetic pick-ups. This preamplifier has RIAA equalization (see Chapter 12 and Fig. 56). (b) Response of the preamplifier (dotted line) compared with the standard RIAA curve (solid line).

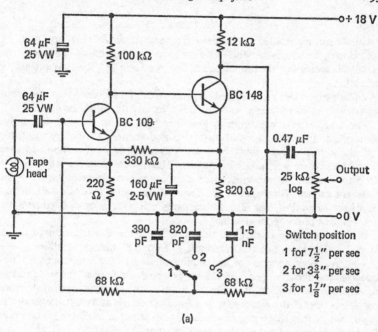

(a)

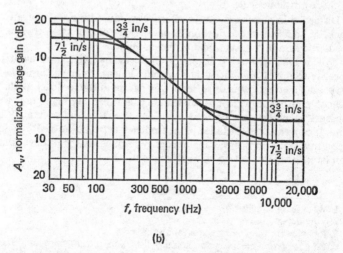

(b)

Fig. 50. A selective feedback amplifier with a response suitable for the replay of magnetic tape recordings: (a) circuit diagram; (b) replay response curve.

Summary

Amplifiers are needed to increase the very small voltages produced by transducers to values suitable for driving power amplifiers. They may be classified according to the function they perform or their frequency of operation.

Voltage amplifiers are designed to produce amplified versions of the input signal without introducing, as far as possible, any distortion or noise.

The basic single-stage voltage amplifier uses a transistor as the active component. In order to avoid distortion and ensure satisfactory transistor operation, bias circuits must be used. Variations of base current are then superimposed on a steady bias current. This produces larger variations of collector current that are superimposed on a steady component known as the quiescent current. Bias circuits are designed not only to obtain satisfactory transistor operation, but also to compensate for variations of temperature and changes of component values.

Where more gain is required than can be obtained using a single stage, several stages are connected in cascade to form a multistage amplifier.

The frequency response of an amplifier tells us over what range of frequencies satisfactory operation can be expected. Because of the wide ranges of responses possible and in order to compare the performances of different amplifiers the relative response is measured in decibels and plotted against the logarithm of the frequency; if suitable graph paper is available we can plot the relative response against the frequency when the 'x-axis' is ruled according to the logarithmic scale.

Distortion in amplifiers can be effectively reduced by using negative feedback. In addition to a reduction in many types of distortion there is also a useful increase in the bandwidth of the amplifier.

Provided the open-loop gain of an amplifier is large, the closed-loop gain of a negative feedback amplifier is almost independent of the performance of the basic amplifier but depends solely on the nature of the feedback path. When the components of the feedback path are close-tolerance high-grade resistors feedback amplifiers can be made with a very stable and accurately known gain. If capacitors are included in the feedback network the feedback depends upon frequency. In this way the frequency response of the feedback amplifier can be tailored to meet specific requirements.

Questions

1. What is an amplifier?
2. Why do we need amplifiers?
3. Describe the amplifying action of a transistor.
4. Why is biasing necessary to minimize distortion in transistor amplifiers?
5. Design an audio amplifier having the circuit of Fig. 40(c). The quiescent current is 1·5 mA and the supply voltage is 15 V. The lowest frequency to be encountered is 60 Hz.
6. Why do all modern high-quality amplifier circuits incorporate negative feedback?
7. Suppose you are considering buying one of two amplifiers. The manufacturer of

one of the amplifiers states that the response of his amplifier is from 30 Hz to 50 kHz; the manufacturer of the second amplifier states that the response of his amplifier is flat within ± 1 dB from 40 Hz to 30 kHz. Which amplifier would you choose and why?

Suggested Further Reading

Ness, H., *Handbook of Tested Transistor Circuits*, Babani Press, 1971.
Olsen, G. H., *A Course Book for Students*, Butterworth, 1973.
Squires, T. L., *Beginners Guide to Electronics*, Newnes–Butterworth, 1975.

5

INTEGRATED CIRCUITS

The invention of integrated circuits is perhaps the most important advance ever made in electronics. There is little doubt that the historian of the future will see the development of silicon device technology as the greatest single factor in the advances of the twentieth century. Economic, reliable and compact systems, from watches and hearing aids to complicated digital computers, are no longer in the realms of science fiction. Without integrated circuits the progress made in space research, supersonic aircraft, communication systems and automation would not have been possible. These tiny devices are now being incorporated in almost every form of modern machinery. Integrated circuits are rapidly becoming the key to measurement and control, which are the bases of all technology.

There are two types or groups of integrated circuit. The first group is often referred to as containing the so-called **hybrid integrated circuits**. These hybrid circuits are a natural development following the use of printed circuit boards instead of wire to effect the interconnections between the various passive and active components. In one form of hybrid circuit the passive components and interconnections are produced by the deposition of thin films in a vacuum. The films are formed on insulating substrates consisting usually of small, flat, thin wafers of alumina. The active devices (transistors and diodes) are subsequently bonded on to the circuit using ultrasonic techniques.

The other form of hybrid circuit consists of **printing** thick films (about 1 thousandth of an inch in thickness) on to an alumina substrate. The printing process is a refined development of the 'silk screen' method used to make advertising posters. After printing, the circuit is 'fired' in a furnace. This firing melts the carrier glaze in the ink, which sets hard on cooling. Research into forming 'printed' transistors has not proved successful and so, as with the thin-film hybrid circuit, we can only print the passive components and connectors; the transistors have to be added at a later stage. These transistors, known as **flip-chip transistors**, consist merely of the transistor chip itself; the plastic or metal container and filling are absent. The tiny transistor has no leads, but has three small connector pads making contact with the base, emitter and collector respectively. The small chips are held by suction to a bonding rod which vibrates at ultrasonic frequencies. The ultrasonic vibrations cause localized heating which welds the transistor contacts to the appropriate points on the circuit.

The Silicon Integrated Circuit

The most important type of integrated circuit is the **monolithic** variety—monolithic because the entire circuit is fabricated within a single block of silicon crystal. Many millions of this type of circuit are sold throughout the world each year. The circuit designers have now progressed so far that almost every type of electronic circuit can be bought as a system within a single package. Complete amplifiers as well as timers, logic gates, memories and computer systems for pocket calculators, clocks and watches can all be bought in single packages. The most common containers are the **TO-5** type and the **dual-in-line (dil)** type.

The **TO-5** is a common metal can, used frequently for packaging single transistors. The can consists of a small cylinder and the lead wires are brought out in the same way as for a transistor. Naturally, there are many more lead wires; usually eight leads arranged in a circular pattern are used.

The **dual-in-line** type (which is much more common) consists of a plastic bar in which the integrated circuit is embedded. The bar is often almost 20 mm long, about 5 mm wide and barely 2·5 mm thick. The leads are brought out along the two sides, hence the term 'dual-in-line'. For most integrated circuits there are seven or eight leads along each side, making either a 14-pin dil or a 16-pin dil package. For clock chips, calculator chips and the more complicated systems 22 or 24 leads are common. Fig. 51 shows diagrams of the common types of package.

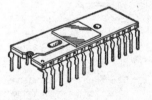

(a) TO-5 metal can   (b) 14-pin plastic dual-in-line (DIL)   (c) Multiple pin chip package for calculators and digital clocks

Fig. 51. Integrated circuit packages.

The manufacture of these circuits is an extension of the planar epitaxial process described in Chapter 3 in connection with the fabrication of transistors. When discrete transistors are made by this method large numbers of identical transistors are made on a single slice. Subsequently the slices are cut and the individual transistors mounted in their separate packages. In order to build a circuit it is necessary to reconnect the transistors electrically and incorporate the passive components. It is not difficult to see how the idea was conceived of building the entire circuit on a single chip. By doing so the separation and subsequent reconnections of the transistors can be avoided.

We have already discussed the planar epitaxial technique when describing the fabrication of diodes and transistors. The extension of the technique, showing the various steps in the production of an integrated circuit, is shown

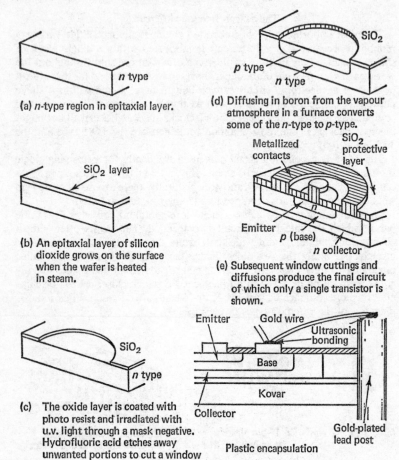

(a) *n*-type region in epitaxial layer.

(b) An epitaxial layer of silicon dioxide grows on the surface when the wafer is heated in steam.

(c) The oxide layer is coated with photo resist and irradiated with u.v. light through a mask negative. Hydrofluoric acid etches away unwanted portions to cut a window

(d) Diffusing in boron from the vapour atmosphere in a furnace converts some of the *n*-type to *p*-type.

(e) Subsequent window cuttings and diffusions produce the final circuit of which only a single transistor is shown.

(f) Stitching of chip to connectors

Fig. 52. Some of the stages in the production of an integrated circuit.

in Fig. 52. Fig. 53 shows how a portion of an electronic circuit is realized in integrated circuit form.

Integrated circuits (i.c.s) may be divided into two classifications: namely, **analogue** and **digital** circuits. Analogue circuits include linear amplifiers in which, over the operating range, the output is a linear function of the input. Digital circuits use transistors as on/off devices, and are used in logic and computer work (see Chapter 13).

In the early days it was widely believed that the application of silicon integrated circuit (s.i.c.) technology to linear circuits would be severely limited by the absence of large capacitors, inductors and good complementary

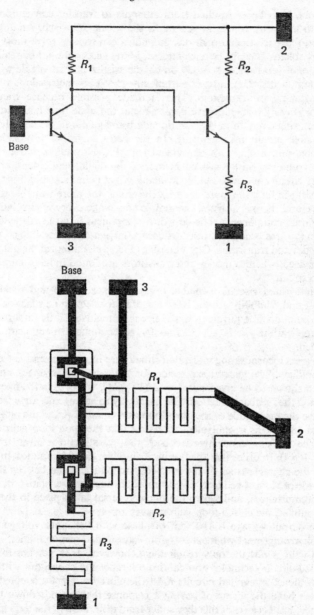

Fig. 53. Realization of portion of a circuit on a silicon chip. The diagram represents a top view of the chip.

transistors. This belief resulted from attempts to transfer conventional circuits into i.c. form, which, in general, is the wrong approach. An approach based on feedback operational d.c. amplifiers can usually solve most of the circuit problems likely to be encountered. Many i.c.s could not be realized in conventional form since they rely on characteristics not attainable with discrete components. The most important integrated characteristic is the very close matching of transistors. This is made possible because transistors fabricated in any one chip have been made at the same time under identical manufacturing conditions. An additional advantage lies in the complexity of the possible circuit arrangements. In the early stages of production, 20 components on a chip was considered complex. Nowadays entire systems such as counters, shift registers, memories for digital equipment, complex amplifier circuits, etc., are readily available. Cost is not a strong function of complexity; in fact, the more complex the circuit the more valuable the chip becomes, and there is a distinct economic advantage over systems built with conventional components. The cost of a component in an i.c. depends very much on the area occupied. Resistors occupy large areas of the chip compared with diodes and transistors. Circuit designs therefore minimize the number of resistors used, whereas diodes and transistors are found in large numbers in any one circuit.

The integrated circuits available today are inexpensive and reliable; in fact, cost and reliability are the basic reasons for selecting i.c.s instead of the discrete component equivalent. Small size is of relatively little importance in many applications—it is a bonus so far as equipment manufacturers are concerned.

Amplifiers incorporating integrated circuits are basically negative-feedback d.c. amplifiers. The general expression for the gain of a feedback amplifier has been shown to be given by $A' = A/(1 - \beta A)$, where A is the open loop gain and β the feedback fraction. The usual way to arrange negative feedback is to have the magnitude of the open-loop gain equal to A and the phase shift equal to 180°. This is equivalent to saying that the open loop gain is $-A$, and hence for simple negative feedback the overall gain is given by $A' = -A/(1 + \beta A)$. In order that the feedback amplifier should be almost independent of the characteristics of the open-loop amplifier, it can be seen that the magnitude of A must be large, and βA must be much greater than one. Under these circumstances, and where the phase reversal takes place in the open-loop amplifier, the closed-loop gain is given by $A' = -1/\beta$, i.e. $|A'|^* = 1/\beta$, and the output voltage is 180° out of phase with the input voltage. For a feedback arrangement we must therefore have a high-gain amplifier.

High gain is not the only requirement, however. It is not possible with integrated-circuit technology to fabricate large-value capacitors with small physical areas. Integrated circuits must therefore be directly coupled. D.C. amplifiers have the virtue of having a response that extends down to zero frequency, but because of this they suffer from drift. Drift is the term given to slowly varying output voltages that exist when the input voltage is zero. It

---

* $|A'|$ means 'the magnitude of A'.

can be caused by varying supply potentials, but the major cause is due to the variations of transistor quiescent current that result from temperature variations. Linear integrated circuits must therefore be high-gain drift-free amplifiers.

The Emitter-Coupled Circuit (Long-Tailed Pair)

One way to reduce drift in d.c. amplifiers is to use the emitter-coupled circuit shown in Fig. 54. Two matched transistors have their emitters con-

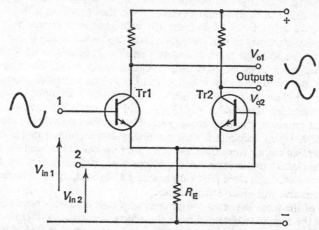

Fig. 54. The emitter-coupled amplifier. Input voltages at terminals 1 and 2 are measured relative to the negative supply line. The waveforms shown refer to the position where $V_{in\,2}$ is steady and $V_{in\,1}$ is a varying signal.

nected to one end of a resistor R_E; the other end of R_E is connected to the negative supply rail. The action of the circuit is as follows:

Let us suppose that $V_{in\,2}$ is held at a fixed potential so as to maintain the correct bias conditions for Tr2. If a signal is now supplied to terminal 1, then rises of $V_{in\,1}$ cause the current in Tr1 to rise and the collector voltage of Tr1, V_{o1}, to fall. A rise of current in Tr1 increases the voltage across R_E. The bias voltage of the base-emitter junction of Tr2 is thereby reduced and the current in Tr2 falls. The voltage at the collector of Tr2 rises. V_{o1} is therefore 180° out of phase with V_{o2}. Provided the circuit is properly designed, the current through R_E is almost constant since rises of current in Tr1 are accompanied by almost equal falls in the current of Tr2. The change in voltage across R_E is therefore only a few millivolts. If signals are applied to terminals 1 and 2 simultaneously then two possibilities exist. If the signals are out of phase we are said to have a **differential** input. Under these circumstances, when the signal at terminal 1 rises the signal at terminal 2 falls. The current in transistor Tr1 increases and in Tr2 the current decreases. The consequence of this is that the collector voltage of T1 falls whilst the collector voltage of Tr2 rises. The output voltages are thus 180° out of phase; the output voltage across the two collectors thus has a magnitude of twice the voltage at any one collector.

Alternatively, if the input signals are in phase then the output voltage at each collector will either rise simultaneously or fall simultaneously. The output voltage across the collectors will then be zero. Input signals that are in phase are said to be **common-mode** signals. If the transistors are perfectly matched no output appears across the collectors. There has thus been a perfect rejection of the common-mode signal inputs. The common-mode rejection ratio (which is the ratio of output voltage to the input voltages applied at either terminal 1 or 2) would be zero. Since the logarithm of 0 is minus infinity, we are unable to describe the common-mode rejection ratio in decibels satisfactorily in this case. In practice, some imbalance in the circuit is inevitable and hence there is usually a small voltage developed across the collectors when common-mode input signals are applied. For integrated circuits, the common-mode rejection ratio is in practice often as much as −90 or −100 dB. For −100 dB this would mean that the output voltage was only 1 hundred thousandth (i.e. $\frac{1}{100000}$) of the input voltage, which, for practical purposes, can be taken as zero. The excellent rejection of the common-mode signal accounts for the popularity and usefulness of the circuit. In integrated circuit form the transistors are physically close and since they were fabricated under identical conditions their characteristics match exceedingly well. Such close physical proximity and matching of characteristics is not possible with discrete transistors.

One of the most important advantages of a high common-mode rejection ratio is the immunity obtained from the effects of variations of temperature and supply voltage. Any effect that increases the current in Tr1 will apply equally to Tr2. The current change in Tr1 and Tr2 will be almost identical. Take, for example, an increase in temperature. Normally this adversely affects the performance of an amplifying circuit. With the long-tailed pair on an integrated circuit chip, however, an increase in temperature will affect both transistors of the pair equally. Since both transistors have almost identical characteristics, the rises in the collector current of each transistor will be the same. The fall of each collector voltage must also be identical and consequently the change in output voltage between the two collectors will be zero. A rise in temperature is in effect equivalent to the application of a common-mode signal. With discrete components, as previously stated, it is impossible to arrange such a close physical proximity, nor is it possible to select transistors with such a close matching of characteristics as we can with an adjacent pair on an integrated circuit. Discrete component emitter-coupled amplifiers are therefore markedly inferior to their integrated circuit counterparts.

The necessary high values of current gains are obtained in integrated circuits by replacing single transistors with Darlington pairs, or even triples. We find also that the performance of the long-tailed pair is improved as the resistance of R_E (Fig. 54) is increased. Large resistances, however, can be obtained only by having long resistor tracks, and these tracks occupy large areas of a given chip. The use of large percentages of the area of a chip is expensive. For this reason the circuit designers try to reduce to a minimum the number of resistors used. Wherever possible transistors are used to provide the necessary resistance value. Transistors do not occupy a large proportion of the chip and therefore are the cheapest component in the circuit. The transistor is not of course used in its amplifying role. By applying fixed bias voltages to the transistor we can make use of the fact that the device has a large dynamic resistance, whilst at the same time not 'absorbing' too high a voltage.

The R_E position of Fig. 54 is an ideal example where the actual resistor can be replaced by a transistor. Since not much voltage is dropped across the transistor the long-tailed pair of amplifying transistors is not starved of voltage. By the time that each amplifying transistor is replaced by a Darlington pair or triple, and R_E is replaced by a transistor equivalent, and various diodes are incorporated for biasing and voltage stabilizing purposes, a single stage of amplification looks quite complicated. Fig. 55 shows how the circuit

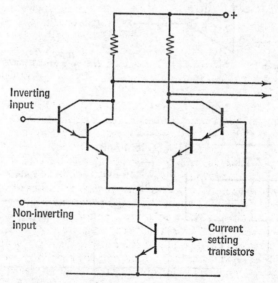

Fig. 55. Input stage of an integrated-circuit linear amplifier. The absence of a circle for the transistor symbols indicates that an integrated circuit is involved.

may look with the biasing arrangements omitted. However, as has already been stated, the transistor is a really cheap component to produce in integrated circuit form, and so the complexity of the circuit does not lead to embarrassingly high production costs.

Integrated Circuit Amplifiers

One of the great advantages of integrated circuits to the builder of electronic equipment is the ease with which complicated systems can be assembled. All the hard design work has already been done by the manufacturers of the integrated circuit. Consider the preamplifiers of Figs. 49 and 50, for example. By replacing the amplifier section with an integrated circuit we can avoid the designing and assembly of the resistors, capacitors and transistors and concentrate only on the system we are trying to produce. The integrated circuit forms the main amplifier section of a feedback amplifier. All we need to do to produce different systems is to alter the feedback circuit. Fig. 56 shows how two preamplifiers can be produced. Fig. 57 shows the circuit of a multimeter that uses an integrated circuit.

The preamplifier circuits of Fig. 56 are not mere replacements for their discrete component counterparts. It should be realized that the open-loop

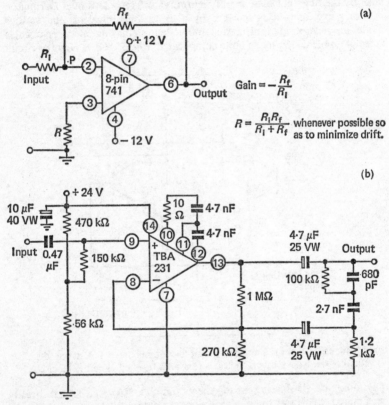

Fig. 56. Two preamplifiers that use integrated circuits instead of discrete components. (a) Basic feedback amplifier. Point P is called a virtual earth because the potential there is never more than a few tens of microvolts from earth or zero potential. (b) A preamplifier for use with magnetic pick-ups. This circuit uses a low-noise integrated circuit TBA 231. Two amplifiers are mounted in the same package. By repeating the above circuit an excellent high-fidelity stereo preamplifier is constructed. The components between pins 10, 11 and 12 ensure stable operation of the amplifier.

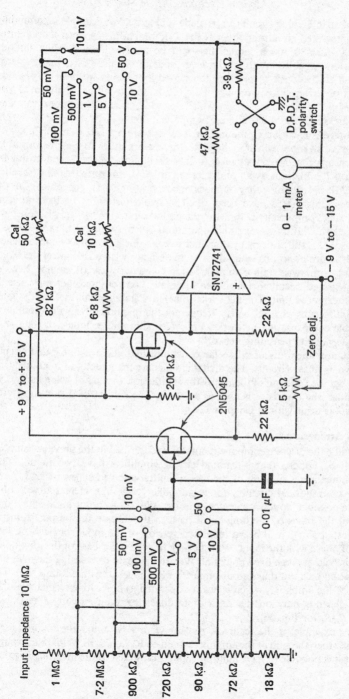

Fig. 57. F.E.T. millivoltmeter.

gain of an integrated circuit can be made very large. For example, a commonly used integrated circuit amplifier is the 741, which has an open-loop gain of 200,000. This is much larger than can be easily obtained when discrete transistors are used. With such a large open-loop gain we can apply a large amount of feedback to produce the closed-loop gains we need in practice. Since the performance in respect of distortion, phase-shift, etc., improves with increasing feedback, we obtain with integrated circuits a superior amplifier compared with the usual discrete component counterpart.

However, when large amounts of feedback are applied to amplifiers with a large open-loop gain it is possible for the amplifier to become unstable. This means that the amplifier goes into oscillation and can then no longer perform its function as an amplifier. In the 741 integrated circuit (labelled SN72741, 741C, or other number incorporating 741, depending on the manufacturer), steps have been taken to avoid instability by incorporating stabilizing components within the integrated circuit. When this is done, the user has the advantage of stable operation, but must accept the lower band-widths that result. For many amplifier circuits this is not of any consequence, but there are occasions when the system designer wishes to control the band-width or otherwise 'tailor' the open-loop characteristics. He cannot then use an integrated circuit that has been internally compensated, but must use an uncompensated unit. The networks to achieve stability can then be designed by him to suit his needs. Very frequently the manufacturer will recommend suitable compensating networks. Fig. 56(b) shows an example of the use of an external compensating network.

The amplifiers discussed so far constitute that class of integrated circuits known as linear circuits. The digital circuits are manufactured in an identical way, but the function of the transistor is different. Instead of being used as an amplifier, the transistor is used as a switch. The subject of digital integrated circuits is dealt with in Chapter 13.

Input Arrangements

Unlike the discrete component amplifiers described in the previous chapter, it will be noticed that integrated circuit amplifiers have two inputs. This advantage arises because of the use of emitter-coupled stages instead of the more conventional circuitry for an amplifier stage. An increased versatility of the possible input arrangements results. We can feed our input signal into one of the bases of the long-tailed pair, in which case the output from the integrated circuit will be an amplified version of the input, but will be 180° out of phase with the input. Whilst doing this the other base of the long-tailed pair is held at some fixed potential. We are then said to be using the inverting input; on a circuit diagram this input is often marked with a minus sign. The gain of the amplifier, which is always operated as a negative feedback amplifier, depends only on the ratio of feedback resistance to input resistance (i.e. R_f/R_1 for Fig. 56).

For example, if the feedback resistor were 1 MΩ and the input resistor 100 kΩ then the gain of the amplifier would be 10, i.e. 20 dB. With this arrangement it is possible to have multiple inputs connected to the inverting terminal and therefore we are able to mix several signals. Fig. 58 shows how this is done and gives the circuit of an audio mixer. Such a circuit is useful in disco or recording work where we wish to combine the signals from, say, a microphone, an electric guitar and a couple of gramophone pick-ups. It is then possible to

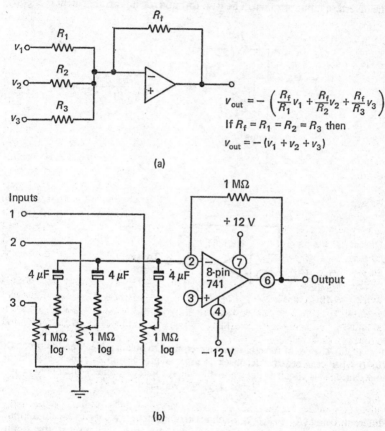

$$v_{out} = -\left(\frac{R_f}{R_1}v_1 + \frac{R_f}{R_2}v_2 + \frac{R_f}{R_3}v_3\right)$$

If $R_f = R_1 = R_2 = R_3$ then

$$v_{out} = -(v_1 + v_2 + v_3)$$

(a)

(b)

Fig. 58. The use of multiple inputs to obtain the addition of input voltages: (a) principle of adding two or more input voltages; (b) circuit useful for mixing inputs from ceramic pick-ups, tuners and guitar preamplifiers.

listen to any one of the inputs individually, swinging from one to another at will; alternatively, we can listen to any combination of the inputs. With this arrangement the input impedance at any input is practically the value of the input resistor. We can thus choose our input resistors to suit the particular transducer, although often we must reach a compromise since we cannot always obtain both the correct input impedance and the required gain simultaneously.

An alternative input arrangement is shown in Fig. 59. Here the signal is fed into the non-inverting terminal so that the output is not only an amplified version of the input, but is also in phase with it. The gain of the amplifier is still controlled by negative feedback, which must of course be applied to the inverting input terminal. The gain formula for this arrangement is slightly

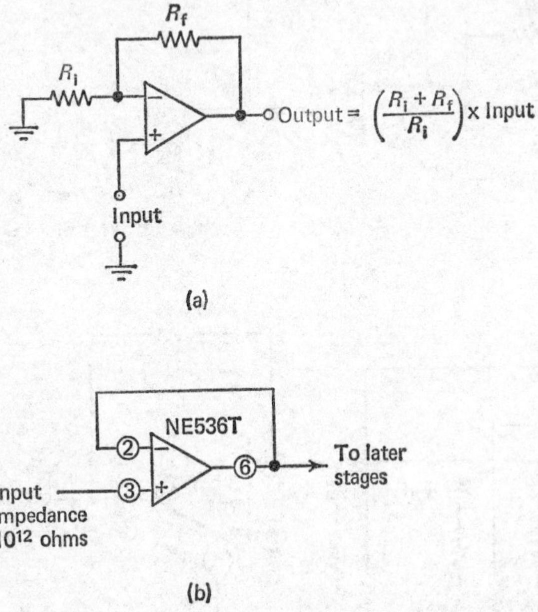

(a)

(b)

Fig. 59. (a) The use of the non-inverting terminal to obtain a high input impedance. (b) 100 per cent negative feedback is used to produce an extremely high input impedance. This circuit is known as the voltage follower, and is often used as the input stage of a voltmeter.

different, being $(R_i + R_f)/R_i$. Such an arrangement gives us a very high input impedance at the non-inverting terminal. We can then adjust the input impedance to our requirements without having to change the components determining the gain. This arrangement is the favoured one for the high-quality reproduction of gramophone records using dynamic pick-ups (Fig. 56(b)).

For voltmeter purposes, where we need the highest input impedance possible (so as not to disturb the circuit under test), we can apply 100 per cent negative feedback by connecting the output directly to the inverting terminal. Here all of the output is fed back to the input so that β, the feedback fraction, is 1. Such a stage does not produce any voltage amplification, but it does have an extremely high input impedance. Stages of this type are known as **voltage followers** and are used as high-input-impedance buffers between the circuit under examination and the amplifying/measuring sections of the voltmeter. Fig. 59 gives an example of a voltmeter input stage with an input impedance as high as 10^{12} ohms, i.e. one million million ohms. A voltmeter with such a high input impedance is useful in areas of scientific work outside of electrical or electronic circuitry. Measuring potentials in living tissues, ionization chambers, in glass electrodes used for pH measurement, in photocells and thermocouples, etc., are typical examples. The extremely high input impedance arises not only from the circuit arrangement used, but also from the fact that conventional bipolar transistors in the first stage of the integrated circuit are replaced by field-effect transistors. In Fig. 59, for example, the Signetics NE536T is used, which is really a 741 integrated circuit with f.e.t. input stages.

Summary

Integrated circuits are perhaps the most important advance ever made in electronics. Whole circuits and systems of circuits can be fabricated by the planar epitaxial technique to produce reliable and inexpensive units of very small size.

Integrated circuits can be classified as **analogue** or **digital**. The analogue type dealt with in this chapter takes the form of a linear operational amplifier. Such amplifiers have a very high open-loop gain and are always operated as negative-feedback amplifiers.

The operational amplifiers being currently manufactured are based on **emitter-coupled transistors**. This arrangement reduces the effect of drift, especially when temperature changes are involved. The use of Darlington pairs ensures that a high gain is obtained. The emitter-coupled circuit also gives us two input terminals, an **inverting input terminal** and a **non-inverting input terminal**. When the **inverting input** is used the ratio of the feedback resistor, R_f, to the input resistance, R_1, is the gain. The input impedance is practically R_1. Because the open-loop gain of an i.c. amplifier is so large, the actual input voltage to the chip at the point P in Fig. 56(a) is never more than a few tens of microvolts. When very large input impedances are required the signal is fed to the **non-inverting** terminal. The feedback is still taken to the inverting terminal so as to be negative feedback. The effective gain then is given by $(R_1 + R_f)/R_1$.

Questions

1. What advantages and disadvantages have integrated circuit amplifiers over discrete component amplifiers?

2. Why was a dual-in-line package devised when it seems that perfectly good cans of TO-5 type were in plentiful supply?

3. Why must an i.c. operational amplifier always be used with negative feedback?

Suggested Further Reading

Hibbert, R. G., *Integrated Circuit Pocket Book*, Newnes–Butterworth, 1973.
Marston, R. M., *110 Operational Amplifier Projects for the Home Constructor*, Newnes–Butterworth, 1975.

6

POWER AMPLIFIERS

The amplifiers considered so far have been designed primarily to give output voltages that preserve the waveform of the input signal, and hence the reduction of distortion has been a major consideration in the design. Although such circuits develop power in their collector loads, this power is of little importance. The choice of transistors and the associated components is not influenced at all by power considerations except in so far as these components have to be operated within their maximum power ratings.

Power amplifiers, however, are those in which the power output is the chief consideration. These are the amplifiers that are designed to operate into loads such as loudspeakers, servomotors, potentiometric recorders, moving-coil pen recorders, etc. The aim is usually to deliver the maximum power into the load consistent with a reasonably low distortion. Unlike voltage amplifiers, which normally operate with comparatively small voltage swings well within the supply voltages, power amplifiers must use all the available supply voltage in order to operate efficiently. With voltage amplifiers the distortion due to non-linearity of the characteristics is not usually a problem because we arrange to operate over only that small part of the characteristic which can be considered a straight line. Since power amplifiers must operate over a much wider swing of voltage, the distortion levels would be unacceptable unless special circuitry is used.

Load Matching

The starting point in the design of a power amplifier must be a consideration of the load, since the object of the exercise is to obtain a circuit that will deliver the required power into the load with the maximum efficiency and minimum distortion. The power requirements of a loudspeaker in a domestic hi-fi system may be as much as 20 or 30 W, whilst for discotheque purposes several hundreds of watts may need to be dissipated to produce the necessary sound levels. For motor drives in servo and recording systems it will be found that many tens or hundreds of watts may be required. These large powers can be obtained only by the use of power transistors capable of handling currents of several amperes rather than milliamps.

The heat produced is often quite large so that the transistors themselves must be a good deal larger than those used for small-signal voltage operation. Furthermore, it is necessary to have these transistors connected thermally to metal cooling fins known as **heat sinks**. The rises in temperature of the power transistors are thus limited to values that ensure that the transistors are not destroyed. The production of heat means that energy is being wasted and so we must arrange that as little heat as possible is generated. There are two

110

ways of doing this: one is by **matching** the amplifier correctly to the load, and the other is by using a special circuit arrangement known as a **Class B push–pull output stage** (see page 112). Let us consider these methods separately.

The power needed to drive the load is not determined by the designer of the electronic equipment; it is fixed by the requirements of the system (e.g. a sound system for the home or factory or perhaps a servo system for the control of machine tools). For a given amplifier, maximum power will be delivered to the load when the output resistance of the amplifier is equal to the load resistance. Since the load resistance is fixed by the system, we must devise methods of altering the amplifier characteristics so that efficient transfer of power occurs; we must **match** the amplifier to a given load.

One way of doing this is to use a transformer. Consider, for example, Fig. 60, which couples an amplifier to an 8 ohm loudspeaker. By comparing Fig. 60 with the single-stage amplifier of, say, Fig. 40(d), we see that the collector resistor has been replaced by the transformer primary. Now the optimum load for the transistor amplifier is several thousand ohms, but the actual loudspeaker load is only 8 ohms. Matching can be achieved by choosing a suitable turns ratio between the primary and secondary coils of the trans-

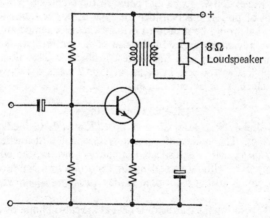

Fig. 60. Circuit of a Class A power amplifier.

former. It turns out that if the load impedance is Z_L ohms and the optimum load for the amplifier is Z_A ohms, then the ratio between the primary and secondary coil turns should be $\sqrt{(Z_A/Z_L)}$. For example, if $Z_A = 8$ kΩ and Z_L is 8Ω, the primary coil should have $\sqrt{1000}$, i.e. about 31·6, turns for every turn on the secondary coil.

Except when low powers are involved (say only 0·2 to 0·8 W), the circuit of Fig. 60 is not often used (see also p. 114). Certainly for higher powers, even though we have the correct matching between the loudspeaker and the amplifier, the efficiency is low. In the absence of any signal the transistor must be biased in such a way that a quiescent current is flowing. If this were not the case several distortion would result. Readers should reconsider the reasons for this given in the chapter on amplifiers. With zero bias, negative-going signals drive the transistor into cut-off conditions, and with too much bias current positive-going signals cause saturation. In both cases much of the signal waveform is cut off. For satisfactory operation, collector current must flow throughout the entire cycle of the signal waveform. The transistor is then said to be operating in the **Class A mode**, i.e. we have a **Class A amplifier**.

Unfortunately, although the distortion of a Class A amplifier is low, the efficiency is also low. This is because a good deal of power is being dissipated by virtue of the standing or quiescent current. For zero input signal, for example, no useful power is being delivered to the load, but the power taken from the supply will be the product of the supply voltage and the quiescent current. The efficiency in this case is zero. When a signal is applied, the useful power delivered to the load increases, but we still have the 'wasted' power to consider because of the quiescent current. Even under the optimum conditions it can be shown that the maximum efficiency of a Class A amplifier with transformer-coupled load is only 50 per cent. Consider a discotheque amplifier that has to deliver 200 W to the loudspeaker. Under the best conditions 400 W would be needed. Of this, 200 W would be delivered to the load, which leaves 200 W to be dissipated in the amplifier. Not only is this very wasteful of power, but consider the type of power transistor and size of heat sink that would be needed to prevent excessive temperature rises.

Another disadvantage arises in the use of a transformer. Not only are transformers costly, they also introduce considerable distortion. For these reasons amplifiers for powers in excess of 1 or 2 watts use what are called Class B transformerless output stages.

Push–Pull Output Stages

We have already seen how Class A operation leads to inefficient power amplifier stages. The quiescent current is essential if distortion is to be avoided. By using more sophisticated circuitry, however, the efficiency can be greatly improved without increasing the distortion by a significant amount. The most common arrangement used is the **Class B push–pull output stage** or its variants.

In Class B operation the transistor is held at the cut-off point by the application of suitable bias arrangements. In the absence of a signal therefore the standing or quiescent current is zero. (For reasons stated a little later in this chapter it is an advantage to operate the transistor at very near the cut-off point and to have a small quiescent current flowing.) Having eliminated the considerable quiescent current of the Class A operation, we increase the efficiency of the stage from a maximum of 50 per cent to a maximum of 78·5 per cent. When an input signal (say of sinusoidal waveform) is applied this signal produces a series of load current pulses that are half sine waves.

It is obvious that it is not possible to use a single transistor as a Class B amplifier because half of the waveform is missing. It is necessary to use two transistors in what is termed a push–pull arrangement. Fig. 61 shows the circuit of such an output stage. Each transistor conducts for half a period and the complete waveform is restored in a special output transformer. This transformer has a centre-tapped primary, the outer ends of which are connected to the collectors of the output transistors, the centre tap being connected to the positive line. Conduction by each transistor for alternate half periods

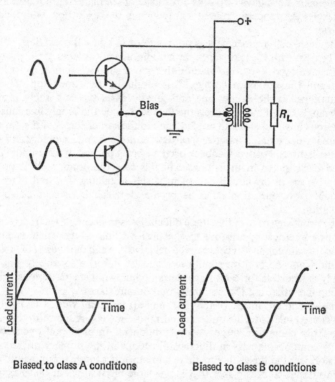

Biased to class A conditions   Biased to class B conditions

Fig. 61. A push–pull amplifier. In Class A the distortion is low and reduces with reducing drive voltages. The distortion with Class B, known as crossover distortion, is due to the curvature of the transfer characteristics near to the cut-off point. For Class B amplifiers the distortion increases with decreasing drive voltages. Many amplifiers are designed for Class AB conditions where a compromise is reached between the efficiency of the Class B mode and the low distortion of the Class A mode.

gives rise to signal flux in the transformer core throughout the period; the complete waveform is therefore available in the secondary output winding.

There are several advantages in using the push–pull type of output circuit. Compared with a single transistor output stage, the same type of transistor in a Class B push–pull circuit delivers twice the power with less distortion. Any second or even harmonic distortion components are cancelled in the transformer because the current components associated with these harmonics are fed into the transformer in phase; the fluxes due to these currents in each half of the primary therefore cancel. Only the wanted fundamental (and odd harmonics) are fed in antiphase to the transformer producing the signal current in the load. The odd harmonics are unfortunately not cancelled in the output transformer. These components are not large, however, and the usual arrangement nowadays is to take advantage of the higher output powers and efficiencies of modern transistors, reducing the distortion with negative feedback.

The push–pull arrangement is not confined to Class B operation, but may also be used with Class A stages. Two additional advantages of the push–pull are then enjoyed. First, the load on the power supply is almost constant in a push–pull Class A output stage. This results in the elimination of voltage fluctuations on the supply line and greatly eases the smoothing problem. With single-ended output stages the fluctuations on the supply line caused by variations in signal strength are fed back to the earlier stages of the amplifier. If care is not exercised, **positive feedback** results, causing instability and spurious oscillations. Positive feedback occurs when a portion of the output voltage is fed back to the input in phase; in this case the product $A\beta$ is positive. Disturbances in the circuit can then be self-sustaining if A and β are large enough. A regeneration effect is produced which is evident as spurious oscillation.

The usual way of avoiding these difficulties is to decouple each stage, which involves interposing resistors in the positive line. Large-value electrolytic capacitors are connected between the negative line and the ends of the resistors remote from the positive supply terminal. In effect, a separately smoothed supply is available for each stage. Decoupling is not such a serious problem in amplifiers that use Class A push–pull output stages.

The second advantage of the push-pull arrangement when Class A operation is used relates to the quiescent currents. In Class A the collector current necessarily contains a large steady component. In push–pull operation the steady components pass in different directions in the primary of the output transformer; the fluxes due to these components cancel. It is possible therefore to design a transformer that is not large physically, but that has a high primary inductance. This results in an improvement of the low-frequency response of the amplifier.

There is no doubt, however, that the output transformer is one of the major sources of distortion and indifferent performance in cheap power amplifiers. Even in high-quality amplifiers it is difficult to attain faithful reproduction of a square-wave signal when transformers are used. A carefully designed output transformer is required that must be physically large so as to have a good response at low frequencies. A happy solution to this problem is possible now that matched power transistors are available. The output impedance of transistor amplifiers can be made very low compared with that of the loads commonly encountered. The use of emitter-follower stages and negative feedback enables transformerless output stages to be designed that have very low output impedances.

Fig. 62 shows the principle of two push–pull circuits. The output transistors together form what is called a complementary pair, because one of them is an *npn* type the other a *pnp* type. When a positive-going signal is applied to the bases, the *npn* transistor (T_1) conducts and the *pnp* transistor (T_2) is cut off. We can see why this happens by considering the base-emitter voltage for each transistor. In the case of T_1 the base voltage is positive with respect to the emitter voltage when the signal is positive-going. For this *npn* transistor therefore the base-emitter junction is forward-biased and so normal transistor action occurs and the loudspeaker forms an emitter load. Effectively then we have an emitter-follower amplifier. Meanwhile, the *pnp* transistor, T_2, is cut off

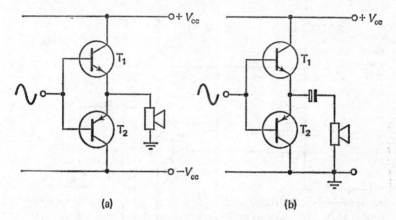

Fig. 62. Complementary push-pull output power stages: (a) uses a centre-tapped supply rail; (b) uses a single supply rail.

because its base-emitter junction is reverse-biased. (We must remember that for a *pnp* transistor the base-emitter junction is reverse-biased when the base is positive with respect to the emitter.) As the signal waveform progresses through its cycle the point will be reached when the signal voltage is negative-going with respect to the zero or earth potential. Under these circumstances the *pnp* transistor becomes conducting whilst the *npn* transistor is cut off. Hence for part of the cycle one transistor is 'pushing' current through the load, and for the remaining portion of the cycle the other transistor is 'pulling' current through the load; hence the term 'push–pull'.

When the transistors are biased to the cut-off point we must remember that they do not conduct fully until the base-emitter junction is forward-biased by at least 600 mV. The push–pull amplifier does not therefore deliver a sinusoidal current to the loudspeaker when a sinusoidal input voltage is applied to the bases. The resulting **crossover distortion** (which occurs during the transition period as one transistor is being switched off whilst the other is being switched on) is unacceptable to hi-fi enthusiasts. The large distortion can be reduced considerably if the transistors are each operated with a small forward-bias so that both are slightly conducting when the input voltage is zero. Operation over the most curved portion of the base-emitter *pn* junction characteristic is thus avoided. The use of negative-feedback further reduces the distortion.

A practical circuit incorporating these features is shown in Fig. 63. The 1·2 volts needed to ensure that both transistors are slightly forward-biased in the absence of a signal is developed across D_1 and R_1. Usually we make R_1 adjustable so as to obtain the lowest distortion consistent with a suitably low

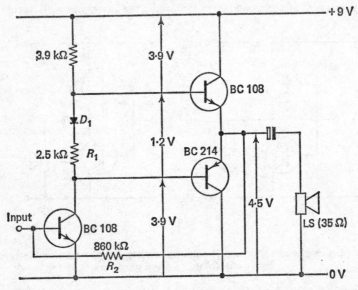

Fig. 63. A practical circuit based on the prototype arrangement of Fig. 62(b). Power output is about 200 mW. Higher powers can be obtained by using different transistors (say a matched pair AD161/AD162), increasing the supply voltage and adjusting R_1 and R_2.

quiescent dissipation. Whilst it is possible to obtain the correct quiescent current by a suitable choice of R_1 alone, such an arrangement would not maintain the correct quiescent conditions as the temperature of the output transistors varied. The inclusion of D_1 introduces a compensating feature since, as the temperature of the transistors increases, so does that of the diode, provided the latter is mounted on the same heat sink. The forward voltage drop across the diode would consequently fall as the temperature increases, hence reducing the quiescent current. This offsets the rise in quiescent current with increases of transistor temperature and thus a stabilizing action is operating. The negative feedback of a proportion of the output voltage is achieved via R_2.

Complementary transistors (*pnp/npn*) are not readily available as matched pairs because of the difficulties inherent in the manufacturing process. It will be realized that matched characteristics of the output transistors are essential in preserving a faithful waveform. If one transistor had a much greater gain than the other, then the portion of the waveform processed by that transistor with the greater gain would be exaggerated because of the larger amplification. One way out of this difficulty is to use two *npn* transistors as shown in Fig. 64.

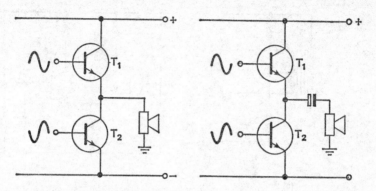

Fig. 64. Quasi-complementary push–pull circuits.

The pair of transistors is not truly complementary (because they are both of the same type) but they can be made to operate in a complementary way by operating them as described below. For this reason they are often referred to as a quasi-complementary pair. With this arrangement it is easy to obtain a matched pair of transistors, but unfortunately some disadvantages arise. One is concerned with phasing.

In order to obtain push–pull operation it is necessary to supply each base with a separate signal; these two signals, although having the same waveform, must be 180° out of phase. Further circuitry is therefore required to provide the necessary pair of anti-phase signals.

The other major disadvantage concerns the mode of operation. When T_1 is conducting this transistor is operating as an emitter-follower; when T_2 is conducting we have common-emitter operation with T_1 acting as the collector load for T_2. Since the two modes of operation are very different, the gains, frequency response input and output impedances of each half of the push–pull output stage need to be 'adjusted' to produce a balanced effect. Once again we rely mainly on negative feedback to overcome the difficulties.

Push–Pull Driver Stages (Phase-Splitters)

The simplest driver stage is the **transformer phase-splitter** of Fig. 65(a). Transformer coupling is quite often used in ordinary radio receivers, but is not really suitable for use in high-grade equipment because of the poor frequency response and phase-shift characteristics of inexpensive transformers. Much better performance can be obtained by using resistor–capacitor elements.

To be satisfactory a phase-splitter should give two outputs of equal amplitude and exactly 180° out of phase. The high-frequency response should be well maintained and, if possible, there should be some useful amplification. Fig. 65(b) shows a conventional emitter-coupled phase-splitter. As the action of this circuit has been described in the chapter on integrated circuits we will not discuss it further here. Fig. 65(c) shows a complementary pair of driver

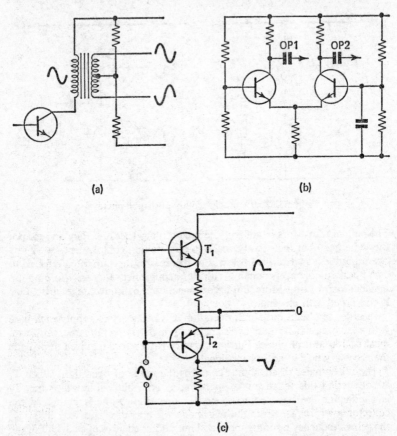

(a)

(b)

(c)

Fig. 65. Various examples of circuits used for driving push–pull output stages: (a) transformer phase splitter; (b) conventional emitter-coupled phase splitter; (c) the use of a complementary pair to provide driving signals for the output stage (biasing arrangements are omitted).

transistors being used as a part of a quasi-complementary output stage. When used with a matched pair of *npn* power transistors they form two Darlington pairs. Referring to Fig. 65(c), when the signal is positive-going transistor T_1 is turned on, whilst transistor T_2 is non-conducting. During the next part of the cycle, when the signal is negative-going, the *npn* transistor T_1 is cut off whilst the *pnp* transistor T_2 is driven into conduction. The outputs from T_1

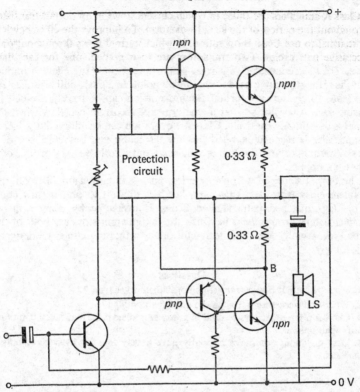

Fig. 66. Simplified circuit diagram of a power amplifier that uses a quasi-complementary pair. If inadvertently the output terminals are shorted, potentials are developed at A and B that activate the protection circuit and apply bias voltages to the bases of the driver transistors in order to prevent the flow of excessive current.

and T_2 are then used to drive the power transistors. Fig. 66 shows a simplified circuit diagram of a power amplifier that incorporates these features. In practice additional circuitry is included to protect the output transistors from damage should a short-circuit arise across the output terminals.

Summary
Power amplifiers are required to drive loads such as loudspeakers, recorders, motors and indicators. They must be able to deliver maximum power into a load, but at the same time introduce the minimum amount of **distortion**. Maximum power cannot be delivered unless the amplifier is correctly **matched** to the load. Although transformers can be used for matching purposes it is now more usual to design suitable transistor circuits that do not require output transformers.

Class A amplifiers are those in which current flows in an output transistor throughout the period of the signal waveform. To improve the efficiency it is more usual to use **Class B** operation, in which current flows throughout only successive half cycles. Two transistors are then needed, one for each half cycle. They are connected in a **push–pull** arrangement. True Class B implies that the transistors are biased to the cut-off point. In a push–pull arrangement this leads to **crossover distortion**. Such distortion can be largely avoided by biasing each transistor so that a small quiescent current can flow when the signal is absent. Alternatively, **Class AB** operation can be adopted in which a compromise is reached between the low distortion and low efficiency of a Class A amplifier and the higher efficiency, but also higher distortion, of a Class B amplifier.

The output stages of a transformerless push–pull power amplifier can use **complementary transistors** (*npn/pnp*), but it is difficult to obtain truly complementary matched pairs. The difficulty is overcome by using a **quasi-complementary** arrangement in which the output transistors are both of the same type, usually *npn*. The push–pull driver stages must then be designed accordingly.

Questions

1. Why are output power transistors larger than other types?
2. When is it necessary to match the amplifier and load?
3. Can coupling capacitors be used in power amplifiers if the load is a d.c. motor? Justify your answer.
4. Why do audio amplifiers normally have a very low, or even zero, output impedance?

7

POWER SUPPLIES

All electronic equipment needs to be energized by means of a power supply. In the great majority of cases the power is delivered to the electronic circuit at a steady or fixed voltage. In the early days of radio and electronics the necessary power was derived from batteries, but the large currents and voltages required for thermionic valves made this source inconvenient. Leclanché type batteries were bulky and expensive, and the lead-acid accumulator required periodic attention. Power supplies were therefore invented that took the necessary power from the mains. The invention of the transistor has, however, brought the battery back into favour. Since transistor apparatus usually requires low currents at low voltages (say up to 24 V) the advantages of small size, cheapness and portability can be realized when batteries are used.

Batteries

Little need be said here about the lead–acid accumulator or Leclanché type cells since they have been extensively described in many textbooks. Most portable transistorized equipment uses a primary battery of which the Leclanché or so-called 'dry' battery is best known (Fig. 67).

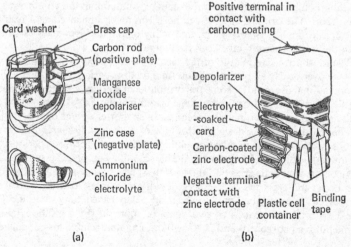

Fig 67. Modern Leclanché cells (Mallory Batteries Ltd. and Electrical Manufacture): (a) single cell; (b) layer type high-voltage battery consisting of several cells connected in series within a single housing.

121

These batteries are available in a wide range of sizes and consist of the appropriate number of cells in series. The common cylindrical cell may be used where comparatively high current discharge rates are required, but for h.t. supplies to hearing aids, apparatus using 'battery' valves, and certain types of transistorized equipment, layer type cells are used. When new, each cell unit has an e.m.f. of about 1·5 V. Such batteries are rated at a definite maximum current discharge rate so that the depolarizing effect will have a chance to keep pace with the hydrogen ion liberation. For batteries of reasonably large volume, the discharge rate may be about 100–250 mA when they are discharged for 100 hours at a rate of about 5 hours in every 24. Under similar conditions, the smaller cells give less current; for example, the familiar U2 cell gives about 30–40 mA for about 100 hours when discharged for 4 hours in every 24. The end-point is reached when the voltage drops to 1·1 V.

Dry cells are intended for intermittent service; they are thus given a chance to recuperate during the rest periods by the action of the depolarizer. This type of cell deteriorates when not in use, the smaller sizes having a shorter shelf-life than the larger. Testing a cell with a voltmeter is of no value when the cell is not delivering current, for even a unit that is almost entirely discharged gives a test reading close to 1·5 V on open circuit. When delivering the maximum rated current the voltage should exceed 1 V; for some applications 1·1 V is the minimum allowed.

Mercury Cells

The mercury type primary cell (for example, those made by Mallory Ltd., Fig. 68) was originally developed during the Second World War for use in portable equipment where maximum energy within minimum volume was the prime aim. The original type of cell has been much improved as a result of research in the last twenty years, and many of the disadvantages associated with the Leclanché cell have been overcome.

The most attractive feature of the mercury cell is the provision of a steady voltage over nearly all of the cell's useful discharge period. Over long periods of operational use, or after some thirty months of storage, a voltage regulation within 1 per cent of the initial voltage is still maintained. Greater degrees of stability and regulation may be obtained over shorter operational periods.

The self-depolarizing design of this type of cell when discharged at current drains within the cell specification eliminates the need for 'rest' periods. For continuous operation of commercial and scientific equipment, transistorized devices, medical apparatus and the like, this proves a distinct advantage over the Leclanché type cell. For emergency alarm devices, rescue radio transceivers, etc., mercury batteries are ideal since they have a long shelf-life. They can be stored for periods of two years or more in dry conditions and at temperatures between 10 and 20°C without any appreciable loss of capacity.

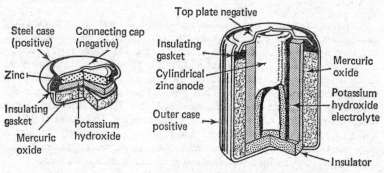

Basic Ruben-Mallory mercury flat cell   Basic Ruben-Mallory cylindrical cell

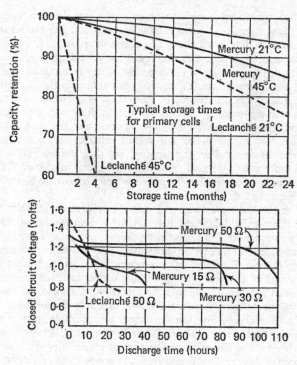

Fig. 68. The basic Ruben Mallory cell with comparative storage times and discharge curves.

(Reproduced by courtesy of Mallory Batteries Ltd. and Electrical Manufacture)

Nickel–Cadmium Cells

Rechargeable nickel–cadmium cells are useful for use in electronic equipment since they can be sealed, thus avoiding the effect of corrosive fumes (which are given off by lead–acid accumulators). The sealed type of cell has a long life (approaching 15 years with care). They can be completely discharged without ill-effects, and can withstand moderate overcharging. The cell's nominal voltage is 1·2 V. Their physical size depends upon their electrical capacity. Button Ni-Cd cells made by DEAC (Great Britain) Ltd. have diameters of 4·3 cm and 5·03 cm with corresponding thicknesses of 0·76 cm and 1 cm for capacities of 450 mAh and 1,000 mAh respectively, to mention only two of their range. This firm also distributes a rectangular cell with a capacity of 7·5 Ah which has dimensions of 4·35 × 5·05 × 10·8 cm.

Solar Cells

The silicon photocell is a photovoltaic device that converts light directly into electrical energy. The more familiar selenium photocell makes this direct conversion, but the efficiency of a selenium cell is too low to allow it to be used as a solar battery. The silicon cell has an efficiency approaching 14 per cent in its present state of development, which is about 25 times greater than that of a selenium cell. (Efficiency in this context is the electrical energy available from the device divided by the solar energy falling on the cell.)

Silicon cells are made by melting purified intrinsic silicon in quartz containers and adding minute traces of a pentavalent element, such as arsenic or phosphorus. The *n*-type silicon that results solidifies and is cut into slices. These slices, after grinding and lapping, are then passed into a diffusion chamber and boron is diffused into the *n*-type crystal from boron trichloride vapour. A *pn* junction results. The *p*- and *n*-type surfaces are then plated and terminal wires added.

It will be recalled from Chapter 3 that a barrier layer, in which very few charge carriers exist, is created between the *p*- and *n*-sides of the crystal, forming a *pn* junction. When discussing the *pn* junction as a rectifier we saw that the application of a reverse bias voltage prevents large numbers of electrons from flowing. The small leakage current that does result is attributed to the production of electron-hole pairs in the barrier layer, the energy coming from thermal sources. In the solar cell there is, of course, no reverse bias voltage, but nevertheless a potential hill exists across the junction. The incidence of radiant energy from the sun creates electron-hole pairs by rupturing the covalent bonds between atoms in the barrier layer. The holes are swept to the *p* side and the electrons are swept to the *n* side (Fig. 69). If an external circuit exists, electrons flow round from the *n* side to the *p* side, dissipating energy in any load that is present. The source of the energy is the incident radiation.

Solar cells are the source of power for energizing the transmitting and other electronic equipment in unmanned satellites. At a less spectacular level they are used as readout devices in computing machines, and in general photovoltaic work. Wherever a source of light is available, solar cells can be used

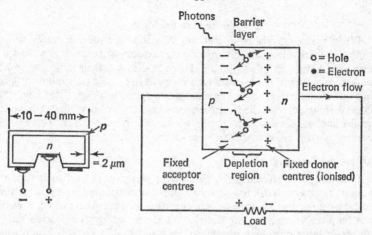

Fig. 69. Principles of operation of a solar cell. The incidence of photons creates electron-hole pairs. The charge carriers are swept out by the field across the barrier layer. An electric current around the external circuit is then obtained.

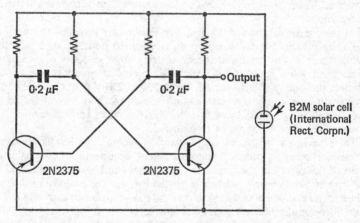

Fig. 70. A multivibrator circuit powered from a solar cell. This type of relaxation oscillator gives an output voltage that consists of square waves.

to energize low-powered transistor equipment instead of the more conventional batteries. Fig. 70, for example, shows a multivibrator being powered by an International Rectifier Corporation solar cell, type B2M. Oscillations are possible with the circuit shown provided the cell is energized with an illumination of at least 10^3 lux ($\approx$ 100 foot-candles).

Power from the Mains

When comparatively large amounts of power are needed, the source of supply is nearly always the alternating current supply mains. For powers not exceeding about ten kilowatts, the **single-phase** supply is usually used. When powers in excess of ten kilowatts are required (e.g. for a transmitter or large industrial equipment), **three-** or **multi-phase** systems are used, together with large mercury or gas-filled rectifiers or some of the heavy power-handling semiconductor devices, such as the silicon-controlled rectifier. These heavy current systems are outside the scope of this book.

Mains Transformers

For reasons of efficiency the bulk of the electrical power needed for the country is generated in a few power stations and distributed via a grid network of cables known as the **mains**. Since enormous powers are involved, it would not be possible to effect the distribution at the voltages considered safe in domestic, laboratory and other locations (i.e. 200–250 V). This would mean that the corresponding currents would be too large to be carried by cables of practical dimensions. The power is therefore generated as alternating current so that it is possible to transform to a large distribution voltage (e.g. 115 kV). For a given power the current is reduced by the same ratio as the voltage is increased. The cables carrying the current can therefore be comparatively thin and cheap.

At the consumer's end of the grid, sub-stations are provided to transform the voltage to a safe value before distribution to domestic and industrial establishments. For industrial locations using large amounts of power, the three-phase 440 V supply is used, but for the type of equipment we are discussing, the supply voltage is from about 220 to 250 V (r.m.s.) single phase. (A single-phase supply is a simple twin-line supply where the voltage on the live line varies sinusoidally with time about a mean earth potential. The neutral line is held at about earth potential.)

The efficient use of this source of electrical energy, like the efficient use of mechanical energy, requires the introduction of some means of converting the form of the energy at the source to a form that can be used by the load. In the case of a motorcar a gearbox is needed between the engine and the road wheels. The gearbox is designed so that the road wheels can turn slowly with great force or alternatively with much greater speeds at less force, depending upon the prevailing conditions. In a similar manner, it is necessary to adjust electrical circuits so that the power available may appear at the load as one of various combinations of voltage and current. The electrical device that corresponds to the gearbox is the **transformer**. It should be noted that neither the gearbox nor the transformer alters the amount of power available if both are 100 per cent efficient. (In practice some loss of power is experienced because of friction in the gearbox and from analogous causes in the transformer.) However, the power is suitably adapted to the particular work to be done.

The operation of a transformer depends upon the principle of electro-magnetic induction. Fundamentally, a transformer consists of two coils that are electrically isolated, but so placed physically that a changing magnetic field set up by an alternating current flowing in one of the coils induces an alternating e.m.f. in the second coil. Thus, mutual inductance is said to exist between the two coils, and the two circuits are said to be **inductively coupled**.

Fig. 71 shows the main physical details together with the diagrammatic representation. The coil connected to the source of power is called the **primary winding**, and the coil connected to the load is the **secondary winding**.

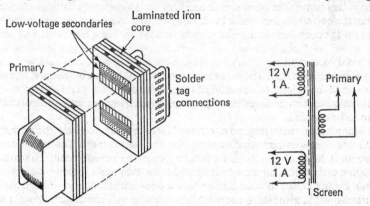

Fig. 71. Construction of a small mains transformer together with the circuit represent-ation of a typical unit used in simple power supplies.

Several secondary windings may be linked with a single primary winding in order to accommodate several different loading conditions simultaneously. The power delivered by the generator passes through the transformer and is delivered to the load, although no electrical connection exists. The connection between the primary and secondary windings is the magnetic flux linkage between the coils. For maximum power transfer all of the lines of flux set up by the primary winding must link the secondaries. To this end, at power frequencies (i.e. 50 Hz in Europe) the coils are wound on a suitable former and adequately insulated from each other. Laminations of magnetically soft iron, or suitable ferromagnetic alloy, are then inserted to form the core.

When no power is taken from the secondary, the supply current and supply voltage are 90°, or $\frac{1}{4}$ or a cycle, out of phase. The power, P, is given by $P = VI\cos\varphi$, where V is the supply r.m.s. voltage, I the r.m.s. current and $\cos\varphi$ is the power factor. When φ, the phase angle, is 90°, $\cos\varphi = 0$. No power is therefore taken from the mains even though the primary is connected to the supply.

In an ideal transformer, when power is being consumed by the load

$$\frac{i_p}{i_s} = \frac{N_s}{N_p} = \frac{e_s}{e_p}$$

where i_p is the primary current, i_s the secondary current, N_s and N_p the number of secondary and primary turns respectively, e_s and e_p the secondary and primary voltages respectively. The output power ($i_s e_s$) is evidently equal to the input power ($i_p e_p$). (Here we are assuming a purely resistive load, i.e. one in which the current and voltage are in phase.)

Practical transformers depart from the ideal in several respects. Not all of the flux induced by the primary is linked with the secondaries; copper losses, due to wire resistance, and iron losses, due to hysteresis and eddy currents, give rise to heat. Copper losses are made good by increasing the number of secondary turns. The open-circuit e.m.f. of the secondary is therefore higher than the output voltage under load. Eddy-current losses are reduced by making up the core from laminations, each lamination being coated on one side by a thin layer of insulating material. Hysteresis losses and flux leakages are reduced by careful selection of the core material and by taking care with the transformer geometry. The core size depends upon the area of the core, A, and also upon the volume to be occupied by the wire of the coils and its associated insulation. Generous core sizes must be employed if undue rises in temperature are to be avoided.

Many transformers have an electrostatic screen wound between the primary coil and the secondary winding supplying the load current. The screen consists of a layer of copper foil extending over the primary coil. The overlapping ends of the foil must be insulated from each other to prevent currents from being induced in what is effectively a one-turn secondary winding. Such currents would give rise to excessive temperatures and consequent damage to the transformer. The purpose of the screen is to prevent mains-borne interference from reaching the secondary circuits; the interference is caused by electric motors, switches, faulty fluorescent lighting equipment, etc. Since such interference is electrostatic in nature the screen must be earthed or connected to the chassis to be effective. Magnetic fields penetrate the copper foil without attenuation and thus transformer action is not affected.

The type of winding used for the supply lines of the electronic equipment will depend on the system of rectification adopted. The specifications for this secondary winding are dealt with in the section on rectifiers.

Variable output voltage transformers (often called variacs, although this is the registered trade name of General Radio) are most useful when it is desired to vary the supply to a piece of apparatus. The device is in effect an autotransformer, i.e. a transformer having a single winding; the required voltage is tapped off in a manner reminiscent of the way in which a potential divider or volume control works. By moving a control knob, any voltage may be selected from zero up to about 5 per cent in excess of the mains. Unlike the potential divider, no power is consumed since the element is almost wholly inductive. It must be emphasized by way of warning that no isolation from the mains is possible, as is the case with a double-wound transformer having an entirely separate secondary winding.

Rectifiers

Because the output voltage of the transformer is alternating, it is necessary to convert it to a direct or steady voltage before it can be used in electronic equipment. The first stage is to convert the a.c. into unidirectional pulses and then these pulses must be passed through a filter, whereupon direct current at a steady voltage is supplied to the load. The process of converting the alternating current to unidirectional current is known as **rectification**, and the filtering after rectification is termed 'smoothing'. Regardless of the type of rectifier used, the function of all rectifiers is the same—that is, they allow electron flow in only one direction.

Metal Rectifiers

Metal rectifiers depend for their action upon the special properties existing at a metal-semiconductor boundary. Two common forms are shown in Fig. 72. The copper/copper oxide rectifier is made by treating copper discs so that a layer of cuprous oxide is formed on one side of the disc. Rectification takes place at the barrier layer between the metal and the semiconducting oxide.

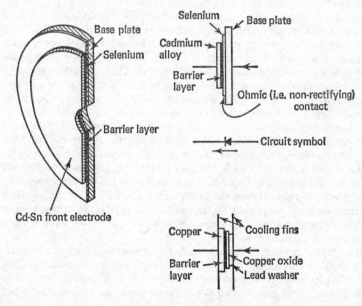

Fig. 72. Forms of power rectifiers of metal-to-semiconductor type. The direction of the arrow shows the direction of conventional current flow when conduction is easy.

The lead disc merely makes good electrical contact between the oxide and the cooling fin, and plays no part in the rectification. The selenium rectifier is made by depositing selenium on an aluminium, or nickel-plated steel disc and applying heat treatment. The selenium is then covered by an alloy of cadmium and tin. The barrier layer is formed by diffusion at the selenium/alloy interface.

Both selenium and cuprous oxide are *p*-type semiconductors. Upon formation of the rectifier cell, electrons diffuse into the semiconductor, thus filling the acceptor centres. In the copper oxide case the electrons come from the copper, which in consequence becomes positively charged. The oxide becomes negatively charged. When the semiconductor is made positive with respect to the copper the potential hill across the barrier layer is reduced and electrons flow freely from the copper to the semiconductor. The rectifier is therefore forward-biased. Upon reversing the polarity the rectifier is reverse-biased and very few electrons can flow. The reverse resistance is about 1,000 times the forward resistance.

In selenium cells it is thought that an n-type cadmium oxide and/or cadmium selenide layer is formed on contact with the selenium. When the p-type selenium is made positive with respect to the cadmium alloy disc, the rectifier is forward-biased. When the conduction is easy the direction of the conventional current (which is opposite to electron flow) is from the baseplate, via a non-rectifying junction to the selenium, and through to the cadmium layer. The reverse-to-forward resistance ratio is about 5,000.

Since the deposits on the metal plates are very thin, the maximum voltage that each type of rectifier can sustain without rupture is quite low. The peak inverse voltage for selenium rectifiers is about 60 V and that for copper oxide types only about 6 V. To overcome this limitation, when higher voltages are to be rectified several discs are combined in series. Thus, the total number of discs determines the peak voltage that can be sustained whilst the effective rectifying area of any one disc determines the maximum current-carrying capacity. The heat which is inevitably generated in the rectifier, must be conducted away from the rectifying discs as quickly as possible. It is usual therefore to interleave each disc with a large cooling fin. All the discs and fins have holes through their centres so that they can be placed in line on an insulating tube. Within this tube is a threaded metal rod. Insulating washers are placed at each end of the stack and nuts at the end of the rod are screwed up, holding all the rectifier elements firmly along the tube. The copper oxide rectifying discs are held tightly on the insulating tube, but the barrier layer of a selenium rectifier must not be compressed unduly, otherwise the resistance in the reverse direction is reduced.

Since copper oxide rectifiers are easily and irreparably damaged by overloads, their use has usually been confined to supplying low current h.t. lines. The selenium rectifier is more tolerant towards overloading, and, when connected as a bridge rectifier, it is often used in battery chargers and other high-current equipment.

Semiconductor Rectifiers

Because of their small size, efficiency and durability, crystal diodes of germanium, or more usually silicon, are now widely used in rectifying circuits. Although the copper-oxide rectifier is still frequently used in moving-coil meters, this and the selenium rectifier are rapidly being superseded by the semiconductor rectifiers. Since the general theory and construction of crystal diodes have been discussed in Chapter 3, we shall limit our discussion here to their practical applications in power rectifying circuits.

Both germanium and silicon rectifiers show marked advantages over the more conventional types. The germanium reverse-to-forward resistance ratio is approximately 4×10^5 and current-carrying capacity is about 50 A/cm² compared with about 0·25 A/cm² for the selenium rectifiers. The peak inverse voltage for germanium rectifiers is often as high as 100 V. The silicon power diode is an improvement on the germanium type. Not only is the maximum operating temperature raised from about 65°C for the germanium type to 150°C for the silicon, but the peak inverse voltage that modern silicon rectifiers will withstand can be as high as 1,000 V per single *pn* junction. Their reverse-to-forward resistance ratio is about 10⁶.

Rectifying Circuits

The **half-wave rectifier** is shown diagrammatically in Fig. 73(a). During one half-cycle, the applied voltage has a certain polarity, and during the succeeding half-cycle the polarity is reversed. Since the rectifier conducts for only one direction of applied voltage, electrons can flow only during half a period. It is for this reason that the arrangement of Fig. 73(a) is known as a

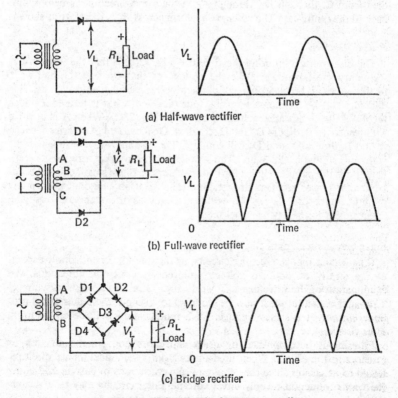

(a) Half-wave rectifier

(b) Full-wave rectifier

(c) Bridge rectifier

Fig. 73. Rectifying circuits for power supplies.

half-wave rectifier. The easy-flow direction refers to conventional positive current, i.e. the flow direction is opposite to that of the actual electron current. The positive sign is usually omitted; on a crystal or metal rectifier the corresponding lead is indicated by a red spot or band on the body of the component next to the appropriate lead. When connected to an alternating voltage, the rectifier polarizes the resistor representing the load (i.e. the equipment being operated) in the way shown.

Full-Wave Rectifier

By the addition of a second diode it is possible to have conduction of electrons in the load throughout the cycle. The arrangement, shown in Fig. 73(b), is then known as a **full-wave rectifier.** If at an instant of time the polarity of A is positive with respect to the centre-tap B, the polarity of C is negative relative to B, and diode D_1 will conduct since it is biased in the forward or conducting direction. The direction of the current is therefore from A through D_1 and R_L and to the centre tap. Half a period later C is positive relative to B, and D_2 is now biased in the forward direction. The direction of the current is then from C, through D_2, through R_L in the same direction as before, and back to the centre-tap. The voltage waveform across R_L is therefore as shown.

Bridge Rectifier

The disadvantage in using the circuit of Fig. 73(b) is that the transformer secondary must produce twice the voltage of that used in the half-wave rectifying circuit because only half of the winding is used at any one time. This difficulty can be overcome by using four diodes in what is termed a **bridge rectifier.** The circuit arrangement is shown in Fig. 73(c). When A is positive with respect to B, diodes D_2 and D_4 conduct. On a reversal of polarity between A and B, diodes D_1 and D_3 will conduct. The resultant waveform across R_L is then as shown in the figure. The arrangement is in effect a full-wave rectifier, and is the popular choice for battery chargers. Since small, efficient semiconductor power rectifiers are readily available at an economic cost, many manufacturers use the bridge rectifier circuit as the standard rectifying arrangement in their power packs.

Filter Circuits for Power Supplies

It is evident that the rectifier circuits, as they stand, do not supply current at the steady or uniform voltage required by electronic apparatus. An examination of the waveforms shown in Fig. 73 reveals a pulsating voltage. The rectified output must now be modified to level out these pulses and produce current at a steady voltage. Such modification is achieved by using **filter circuits.**

Filtering is accomplished by using capacitors and either inductors or resistors in the circuit. When used for this purpose, inductors are often referred to as **chokes** since they choke off any variations of current and allow the easy conduction of only direct current. Filter circuits may take various forms, but the basic arrangements are shown in Fig. 74.

The simplest filter arrangement, shown in Fig. 74(a), consists merely of a capacitor in parallel with the load. The value of capacitance must be large in order to present as small a reactance as possible to the pulsating rectified output, and to store sufficient charge so that current may be maintained in the load during the period that the rectifier is not conducting. The reactance of the capacitor should be much less than the resistance of the load. For the kind

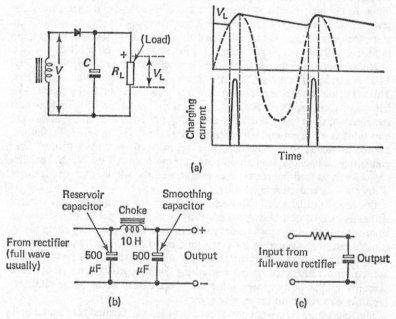

(a)

(b)   (c)

Fig. 74. (a) Approximate waveforms for the half-wave rectifier. (b) A capacitor input filter with typical values of choke and capacitors. The choke presents a very high impedance to current variations and little opposition to d.c. (c) A resistance capacitor filter. The smoothing capacitor provides a very low impedance to alternating voltages, but a very high impedance to direct voltages.

of loads usually encountered, capacitors of 500–1,000 μF are commonly used. The rectified pulses charge the capacitor to a voltage close to the peak value delivered by the rectifier. Because of the large value of C, the time constant CR_L is large compared with the periodic time (0·02 sec) of the applied voltage. The voltage across R_L does not therefore fall sinusoidally, but decays exponentially. The fall of voltage may be reduced for a given load by increasing the value of C. There is a limit, however, to the value of capacitance used. It can be seen that the slower rate of fall of voltage across R_L, the smaller is the time available to recharge the capacitor. The current pulse delivered by the rectifier must therefore have a greater peak value to deliver a given energy. All rectifiers have peak current ratings and these ratings can be exceeded if the value of C is too large, causing damage to the rectifier. For any given rectifier and associated circuit, the maximum value of C that can safely be used is specified by the manufacturer in his published data sheets.

The designer of rectifying circuits must observe an additional precaution. This precaution is concerned with the maximum peak inverse voltage that the rectifier can tolerate. During the time that the rectifier is not conducting, we see from Fig. 74(a) that the peak or maximum inverse voltage applied to the diode is the sum of the voltage across the capacitor and that across the transformer secondary. The peak inverse voltage (p.i.v.) is thus approximately twice the peak voltage across the transformer secondary. The manufacturer's published data give the maximum p.i.v. that may safely be applied.

The output from the rectifier when a single capacitor is used as in Fig. 74(a) is not steady enough for the voltage supply to electronic equipment. The variations would give rise to unwanted components in the signal. For half-wave rectifiers the unwanted component is a 50 Hz one. In full-wave rectification, since two pulses are delivered during each cycle of the mains, the unwanted component has a frequency of 100 Hz. To reduce these variations, known as **ripple**, to a minimum, it is necessary to employ a further filter consisting of a choke and smoothing capacitor as in Fig. 74(b). The first capacitor is termed the **reservoir capacitor,** and the whole filter arrangement is known as a **capacitor input filter.**

For most transistorized equipment involving power amplifiers the current demands can be quite heavy. Frequently it is not economic to use chokes since reasonable inductance values would involve large and bulky components if hundreds of milliamperes (or even amperes) of current are required. Under these circumstances a simple capacitor smoothing is used for the power output section with resistor–capacitor smoothing for the preamplifier stages. The power amplifier output stage must then be designed so as to be relatively insensitive to ripple on the supply lines.

Fig. 75(a) shows a typical power supply which can be used for small amplifiers. Readers should note some points of practical importance. The mains on/off switch should always be a double-pole switch since it cannot be guaranteed which mains input line will be 'live'. Fuses of adequate rating must then follow so that if components in the power pack fail excessive current cannot then be taken from the mains; in the absence of fuses there is always a risk of fire. It is a good idea to have an indicator on the equipment to show the presence of the mains.

Fig. 75(b) shows a power supply suitable for power amplifiers. As previously stated the output stages of the amplifier are designed so as not to be too critical of ripple on the d.c. supply line. As we have seen when describing power amplifiers, it is often necessary to have both a negative as well as a positive supply rail. Conventional resistor–capacitor smoothing is then used as shown, to supply the more ripple-sensitive pre-amplifier section.

Zener Diode Stabilizers

There are numerous occasions when it is necessary to use a power supply that gives a steady output voltage that is not affected by mains voltage

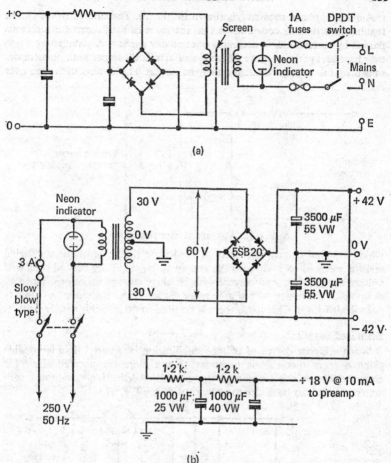

Fig. 75. Two examples of power supply circuits: (a) typical power supply for small currents; (b) a power supply suitable for an audio power amplifier.

fluctuations or load current variations. **Zener diodes** can be used to provide simple voltage stabilization.

The main characteristics and operation of the device have already been discussed in Chapter 3. By applying sufficient reverse bias voltage, an avalanche condition is reached at which point there is a sudden and substantial increase in the reverse current. Thereafter very small increases in reverse bias voltage cause large increases of current. In effect, large current variations are possible, the voltage across the device remaining almost constant. In typical silicon diodes the current may vary over a range of 5–200 mA for a voltage variation of one or two hundred millivolts.

A simple voltage regulator is shown in Fig. 76. The sum of the load and regulator currents is constant, so that increases of load current are accompanied by corresponding decreases of regulator current. A given zener diode may have an operating range of 2 mA to 200 mA, the upper limit being determined by the wattage dissipation of the device. This is also the range over

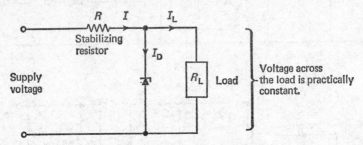

Fig. 76. A simple voltage regulator.

which the load current may vary. To calculate the value of the stabilizing resistor we need to know: (a) the supply voltage; (b) the desired stabilized voltage; and (c) the load current with its likely current variations. Since an example of the design of a simple zener diode voltage stabilizer has already been dealt with in Chapter 3, there is no need to go into this matter further.

Stabilized Supplies

When a higher degree of voltage stabilization is required than is possible using a zener diode, it is necessary to use more complicated electronic circuitry. The principle of a regulator capable of delivering current at a very stable voltage can be understood by considering Fig. 77.

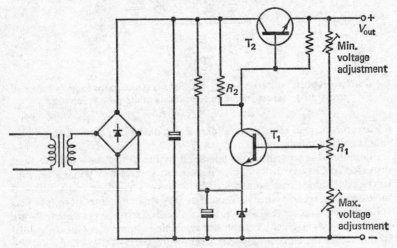

Fig. 77. Stabilized power supply.

If any change occurs in the output voltage, V_{out}, then a proportion of that change affects the voltage at the base of the amplifier transistor T_1. Let us say that V_{out} rises for some reason. A proportion of that rise, determined by the relative values of the resistances in the resistor chain across the output supply, is applied to the base of T_1. Now the emitter of T_1 is held at a constant voltage by the zener diode. There is therefore an increase in the base-collector voltage of T_1. Normal amplifying action takes place, so, because of the rise in base-emitter voltage, the resulting rise in collector current produces a fall of voltage at the collector. The fall in collector voltage is, of course, much larger than the change in base-emitter voltage. The collector of T_1 is connected directly to the base of T_2, which is a power transistor that can handle the whole of the load current. Because the voltage at the base of T_2 falls, the current through T_2 is reduced; this results in an increase of the voltage between the collector and emitter of this ballast transistor, T_2. The rise in voltage across T_2 means that there is less output voltage available. In effect then the increase of voltage across T_2 largely nullifies the rise that took place in V_{out}. We have in effect a negative feedback system that stabilizes the output voltage. It will be noticed that T_2 is being operated in the emitter-follower mode.

Fortunately for the majority of those wishing to build stabilized power supplies, there is no need to be involved in complicated design problems. Regulators working on the principle of the circuit of Fig. 77 are now readily obtainable in integrated circuit form. The circuits available are more sophisticated than the one already described in that high-gain emitter-coupled amplifiers are used with temperature-compensated circuitry, and fairly elaborate arrangements are made to give adequate short-circuit protection. The regulators are therefore almost foolproof. Fig. 78 gives just two examples of stabilized supplies that incorporate regulators from a large range of available units.

The Thyristor

The thyristor is an 'on'/'off' device. Once in the 'on' state, a thyristor remains in this condition so long as there is a substantial current. The control electrode, known as the gate, can only initiate firing; it cannot stop conduction because it loses all control once the main current is started. The thyristor is a high-current, low-voltage device that is triggered by a small current from a signal source. The resistance of a thyristor in the forward or conducting state is only a very small fraction of an ohm; the voltage drop across the device is therefore much lower than can be achieved in thyratrons. Like the transistor, the thyristor is immediately ready for action; it requires no warming-up period. Physically the device is much smaller and more robust than its thermionic counterpart. Powers of tens of kilowatts can easily be controlled by a comparatively small thyristor.

The main constructional features of a thyristor are shown in Fig. 79; it is a three-terminal device consisting of four layers of semiconducting material in a *pnpn* sandwich. There is crystal continuity throughout the structure, and therefore three *pn* junctions exist between the four layers.

When connected in series with a load resistance and d.c. source, current is

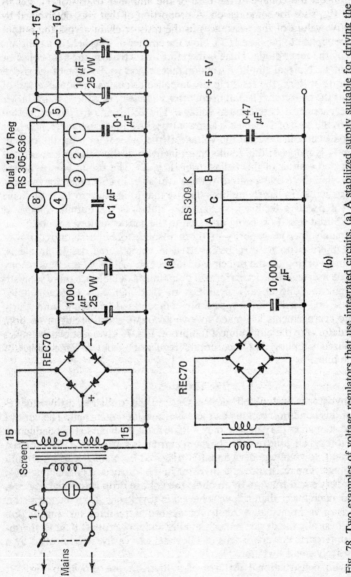

Fig. 78. Two examples of voltage regulators that use integrated circuits. (a) A stabilized supply suitable for driving the majority of linear integrated-circuit amplifiers (e.g. the 741 amplifier). (b) A regulated supply suitable for logic integrated circuits used in digital work.

prevented from flowing in the direction which makes the anode negative. Upon reversing the polarity of the supply voltage, so that the anode is now positive, the device remains non-conducting unless, at the same time, the gate electrode is driven sufficiently positive to produce a gate current that exceeds

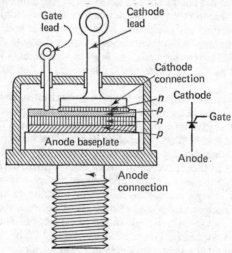

Fig. 79. Constructional features and circuit symbol of a thyristor.

a certain critical value. As soon as the main current is established the gate loses control, and the anode current is then determined solely by the resistance of the external circuit and the value of the anode supply voltage. The current can be cut off only by switching off or reversing the anode voltage, or by introducing so much resistance into the circuit that the anode current falls below a certain maintaining threshold value.

To understand the mechanism, let us consider Fig. 80. When the applied

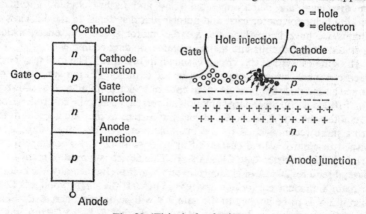

Fig. 80. 'Firing' of a thyristor.

voltage is such that the anode is negative and the cathode positive both the anode junction and cathode junction are reverse-biased. The thyristor is then in the blocked condition and its resistance is very high (about 0.2 MΩ). When the controlled rectifier is subject to a forward voltage (i.e. anode positive) then, in the absence of any gate voltage, the gate junction is reverse-biased and conduction through the device is still prevented. In this condition the majority carriers are withdrawn from the cathode and anode junctions and thus all the applied voltage appears across the gate junction. At this junction the field across the depletion layer is high because the voltages that in practice are applied to the anode are large (e.g. 20 or 30 V up to several hundred volts).

To 'fire' the rectifier a pulse of positive current must be injected into the gate. The holes injected by the gate electrode neutralize the negative charge on the p side of the gate junction. Because of the presence of a positive space charge on the n side of the gate junction, the injected holes migrate along the junction. Although the gate current pulse is relatively small (10 mA of gate current may be all that is needed to switch 20 A of main current), the density of holes between the gate and cathode is high enough not only to neutralize the local space charge, but also to induce the injection of a large number of compensating electrons from the cathode. These electrons are attracted to the positive space charge on the n side of the gate junction. They gain sufficient energy to cause avalanche breakdown (see page 52). This avalanche breakdown spreads right along the gate junction. The controlled rectifier now carries a large forward current. The gate can now no longer influence the conduction process. The rectifier can return to its non-conducting state only when the main carrier density is reduced below the critical value necessary for avalanche multiplication.

Practical Thyristor Circuits

Circuits that incorporate thyristors for control purposes fall into two classes: those using d.c. supplies and those using a.c. supplies. Examples in the first class are battery-operated relay and flasher circuits, automatic parking lights for motor cars, and burglar alarm systems. In the a.c. class of circuits we have lamp dimmers, a.c./d.c. motor speed controllers, mains-operated flasher circuits and mains-operated welding equipment.

The simplest type of d.c. circuit is shown in Fig. 81(a). The load may be an electric furnace, a lamp, a relay coil, or a d.c. motor. (In the latter case this simple circuit does not provide speed control.) Such a circuit as that shown in Fig. 81(a) is useful when large amounts of current need to be controlled from a remote position. R_1 is chosen so that the current to the gate is equal to the firing pulse recommended by the manufacturer. For example, suppose we wished to control a load current of up to 3 A at 12 V, we could select an S.T.C. (ITT) thyristor type CRS3/05AF. This device is typical of many of its class; it requires 20 mA of gate current at 3 V to fire. (Incidentally, the holding or maintaining current for this device is 25 mA.) If the supply voltage is 12 V, then if 3 V is to be applied to the gate 9 V will be dropped across R_1. Since the gate current fed via R_1 must be 20 mA we see that $R_1 = 9$ V/20 mA = 450 ohms. In practice the nearest preferred value of 470 ohms would be used. The presence of R_2 is to prevent the inadvertent firing of the thyristor. If the gate is connected to the switch via a long line it is easy to pick up static or other interference signals on the line when the switch is open. To prevent this, R_2 provides a suitable leakage path. R_2 must have a fairly high resistance compared with R_1 because together they form a potential divider. We do not

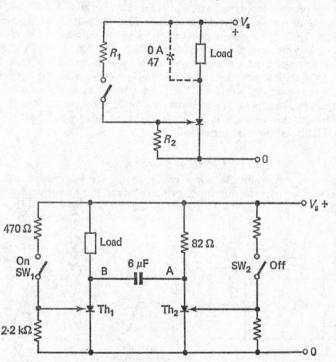

Fig. 81. Two thyristor circuits for controlling current when steady supply voltages are involved. (a) A simple circuit for controlling large amounts of direct current. (b) Use of a second thyristor to turn off the main one.

want the current delivered via R_1 to be diverted from the gate of the thyristor via a low R_2. If R_1 is 470Ω, a suitable value for R_2 would be 2·2 kΩ. If the load happens to be a relay coil, bell or buzzer, it will be necessary to add the diode OA47 as shown. This diode prevents any high voltage transients from developing when the coil circuit is broken. These transient voltages could easily damage the thyristor if they were not damped out by the diode.

Switching off the thyristor in the simple circuit of Fig. 81(a) presents some difficulty. We must either interrupt the supply voltage or somehow reduce the load current below the maintaining current. The beauty of remote control will be lost if we choose to interrupt the supply voltage with a switch because such a switch will have to interrupt the large load current. This would present as much difficulty as starting the large load current. However, we can overcome the difficulty by employing the circuit of Fig. 81(b). Light reed, or other switches, can then be used both for turning on and turning off purposes. The turn-on of the load is effected by closing SW_1 momentarily and thus firing Th_1. The $6\mu F$ capacitor will now charge because one terminal (A) will be at the positive supply voltage, V_s, whilst the other terminal (B) will be at almost the zero potential level because the voltage drop across the conducting thyristor Th_1 will be low. If now SW_2 is closed momentarily, thyristor Th_2 will fire. The potential at A will suddenly fall to almost zero. Since the charge on the capacitor cannot alter instantaneously the voltage cannot alter either ($V = CQ$). So if B is initially at almost zero potential when the potential at A falls to practically zero volts, the potential at B must fall to a negative voltage. Since this point is also connected to the anode of thyristor Th_1 the potential at this anode will momentarily be negative with respect to the cathode; Th_1 is therefore turned off.

Control of A.C. Supplies

Since power from the mains is so readily available it is attractive to control the mains current with thyristor circuitry; indeed the majority of thyristor control gear is associated with a.c. supplies. There are two methods by which

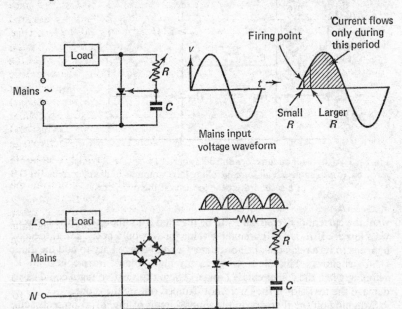

Fig. 82. The thyristor control of alternating current: (a) basic firing circuit for a.c control; (b) full-wave a.c. control.

such control can be effected; these are known as the 'phase-shift' method and the 'burst-fire' method.

In the **phase-shift** method the current flow is restricted to only a portion of the time during every positive half cycle. This is achieved by delaying the firing of the thyristor during each positive half cycle and so preventing the current from flowing throughout the time that the supply voltage is positive-going. The point on the positive-going waveform at which firing commences is controlled by varying the time-constant of a simple CR circuit. Fig. 82 shows the principle. As soon as the mains waveform is positive-going C commences to charge via R. The voltage rise across C is determined by the time constant, i.e. the product CR. It is arranged that for a time the voltage across C is insufficient to fire the thyristor; no current then can flow through the load. At some point during the half-cycle the voltage across C does reach a value sufficient to fire the thyristor. Once fired the thyristor continues to conduct, and power is supplied to the load. At the end of the half-cycle the voltage across the load-thyristor combination falls to zero and the thyristor again becomes non-conducting. The firing point can be controlled by varying R and keeping C fixed at a suitable value.

Such half-wave control is obviously of very limited value since no power can be delivered during the negative-going half-cycles. An improvement can be achieved by first rectifying the mains with a full-wave rectifier. This in effect makes both half-cycles positive-going so control can be exercised twice in a complete mains cycle. During any half-cycle, once the thyristor has fired the bridge is effectively shorted out and current then flows in the load (Fig. 82). A useful practical circuit based on this principle is shown in Fig. 83. It is a circuit for a battery charger with automatic overload protection. Most of the circuit should easily be understood from previous descriptions. The automatic overload protection is provided by a zener diode. By an appropriate setting of R1, a fraction of the desired final battery voltage (equal to the zener voltage) is available at the slider terminal. For all battery voltages that make the slider potential less than the zener voltage, the zener diode does not conduct and SCR2 is not fired. As soon as overcharging commences, the battery voltage becomes higher than the final desired voltage; the zener diode then conducts and SCR2 fires. (SCR = silicon controlled rectifier; we now call the device a thyristor.) SCR2 conducts during each successive half-cycle pulse from the bridge rectifier. The current through R2 is then large enough to cause a substantial voltage drop across this resistor, and the potential at A never rises to a sufficiently high value for SCR1 to fire. If the battery voltage falls the zener diode no longer conducts; SCR2 cannot then be fired. The voltage at A is then high enough to allow SCR1 to fire and charging of the battery recommences.

One disadvantage in using a simple CR circuit to control the firing time as in Fig. 82 is that the control is limited. In particular we must not allow R to become so small that an excessive gate current could flow, otherwise the thyristor would be destroyed. To avoid this difficulty, and give a wider range of delay times, a **unijunction transistor** (U.J.T.) may be used to fire the thyristor. The unijunction transistor fires the thyristor by delivering to the gate

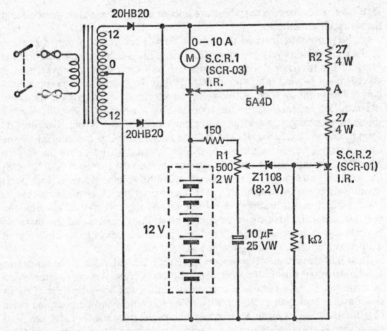

Fig. 83. A battery charger with automatic overload protection. All the diodes and rectifiers are by International Rectifier Corporation, but any similar adequately rated components will be satisfactory. The thyristors are labelled S.C.R.1 and S.C.R.2. A single 12 volt secondary and standard bridge rectifier can be used instead of the two-diode full-wave rectifier shown.

short spikes or pulses of current. The use of pulses is the most satisfactory and reliable method of firing a thyristor.

A lamp dimmer that uses this principle is shown in Fig. 84. We have already discussed in Chapter 3 the way in which a unijunction transistor works. It will be recalled that when the emitter voltage is less than a certain fraction (β) of the voltage between the two bases, the transistor acts as though it were just a resistor. No current enters via the emitter lead because the *pn* junction between the *p*-type emitter and *n*-type bar is reverse-biased. However, when the emitter voltage exceeds β times the voltage between the two bases this junction becomes forward-biased. A substantial increase in charge carriers takes place in the region between the emitter and base 1; this region therefore becomes an excellent conductor, i.e. the resistance between the emitter and base 1 falls to a very low value.

Consider now the action of the circuit for the lamp dimmer (Fig. 84(a)). As the supply voltage from the thyristor rises, the voltage across the unijunction transistor from base 1 to base 2 rises. The voltage to the emitter also rises, but with a phase delay determined by the values of C and R. For any desired firing point in the half-cycle it is possible to adjust R to give the correct phase

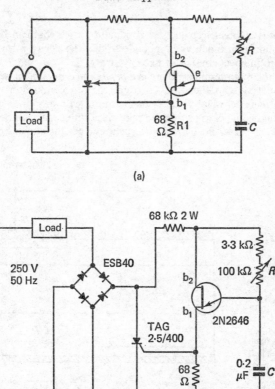

(a)

(b)

Fig. 84. Use of a unijunction transistor oscillator to provide firing pulses for a thyristor. (a) Delayed firing by pulses from a unijunction transistor. (b) Practical version of (a) often used as a lamp dimmer.

displacement and voltage across C so that at the desired time the critical voltage is reached at the emitter of the U.J.T. The rapid fall in resistance between the emitter and base 1 results in the rapid discharge of C via the U.J.T. and the resistor in the base 1 lead. The pulse of current through the 68 ohm resistor gives a short pulse of voltage across the resistor and hence to the gate of the thyristor, which then fires. Once the half-cycle of supply voltage is over, the voltage across the thyristor/unijunction transistor combination becomes zero and the circuit is then ready for the next half-cycle.

The Triac

The triac is a three-electrode semiconductor switch that can be triggered into the conduction mode via a gate electrode. The device is an integrated circuit which in effect consists of two thyristors.

Between the main terminals there is a *pnpn* switch in parallel with an *npnp* switch. The triac can therefore be triggered into conduction with either polarity of main terminal voltage; furthermore, with any given polarity of main terminal voltage, triggering can take place with positive or negative gate signals. Fig. 85 shows the basic triac structure. Once the latching current has

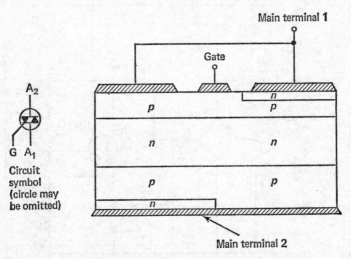

Fig. 85. Basic schematic structure of a triac.

been exceeded the gate electrode loses control. The turn-on time for modern devices is typically $1\mu s$. In order to turn the device off the current must fall below a minimum holding current. Since the latching current is greater than the holding current a hysteresis effect is evident; the gate-drive signal must increase the principal current to a value exceeding the latching current before latching action occurs. Once on, however, the principal current must be reduced below the latching current (i.e. to the holding current) before the device is turned off. The usual way to reduce the principal current to a value below the holding current is to lower the terminal voltage to zero. The turn-off time for recently manufactured devices is typically $50\ \mu s$.

The triac was specifically developed to provide improved means of controlling a.c. power compared with methods involving thyristors; the device offers simplified circuitry to perform the same function. In all cases it is the triggering circuitry that is simplified, which leads to a reduction in the number of components required. Triggering is usually performed by a two-electrode device called a **diac**. This takes the form of a glass encapsulated three-layer device designed specifically for symmetrical positive and negative breakover voltages. Readers will recall that the breakover voltage for a zener diode is controlled during manufacture, but only for negative applied voltages; in the forward direction normal conduction occurs for applied voltages exceeding only 600 mV or so. Diacs have the property of not breaking over for forward applied voltages until a specific value has been reached.

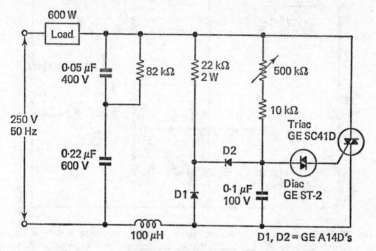

Fig. 86. Complete incandescent lamp dimmer with R.F. filter.

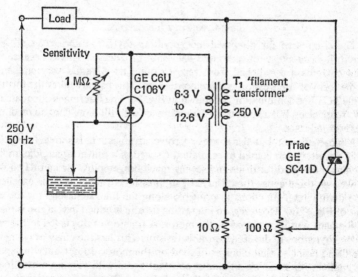

Fig. 87. Water-level sensing control circuit.

Examples of circuits involving triacs are shown in Figs. 86–88; they are taken from Application Note 200.35 published by General Electric and entitled *Using the Triac for the Control of AC Power*, by J. H. Galloway.

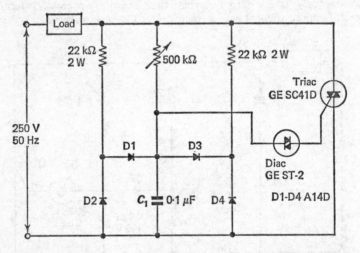

Fig. 88. Wide-ranging hysteresis-free phase control. Loads of up to 600 W may be accommodated with the GE SC41D triac; for greater loads a larger triac will be required. The circuit shown is capable of firing a wide range of triacs.

Radio-Frequency Interference

The circuits so far discussed incorporating thyristors and triacs all give rise to interference with radio and television broadcasts. This is because the load currents are switched from zero to quite large values very suddenly. Some form of suppression is therefore needed. This usually takes the form of some *LC* filter combination, as shown in Fig. 86. The filtering is adequate for TV frequencies, but a good deal of interference will still be evident on medium-wave broadcasts.

A more efficient, but also more expensive, arrangement is shown in Fig. 89. This circuit, which is used to regulate the speed of a universal a.c./d.c. motor such as those used for drills and sewing machines, employs two 4 mH chokes as well as three capacitors. The R.F. suppression of this circuit is excellent. The main circuit functions in principle along the lines described for the circuit of Fig. 82(a). However, an interesting feature is added in that use is made of the back e.m.f. produced by the motor. A fraction of this is fed to the gate of the thyristor via the 500Ω variable resistor. This has the effect of changing the firing point so that changes of load on the motor do not alter the speed very much. Maintaining an almost constant speed with varying motor loads can be a valuable feature, especially in drilling and sawing operations.

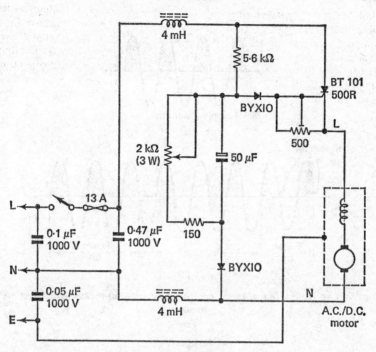

Fig. 89. Variable power supply for controlling the speed of small a.c./d.c. motors.

The Burst-Fire Method

The method of using *LC* filters to suppress radio-frequency interference cannot be economically applied when very large load currents are involved. A typical example would be in furnace control circuitry. The burst-fire control method can overcome the difficulty by switching on the thyristor only at the time that the alternating supply voltage is going through the zero voltage point. The method involves turning on the thyristor for an integral number of half-cycles and then turning the thyristor off, both the turn-on and turn-off actions taking place at the zero main voltage points. The circuitry for doing this is rather too complicated to discuss here, but readers who wish to use the method should know that integrated circuits are readily available to perform the function. Any of the large integrated-circuit manufacturers will be glad to supply details of their devices, known as zero point switches. Motorola Application Note AN-453 entitled *Zero Point Switching Techniques*, available from Jermyn Industries, is a typical example of literature useful to those who wish to study the subject further.

The principle of 'burst-fire' control, compared with the 'phase-delay' method, is shown in Fig. 90.

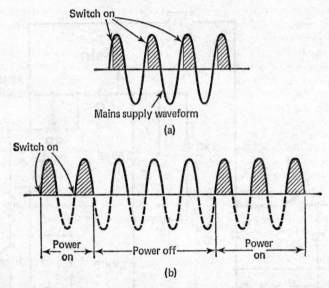

Fig. 90. Two basic forms of power control. (a) Phase-shift method. (b) 'Burst-fire' method.

Summary

All forms of electronic equipment need to be energized by power supplied at a steady voltage. For portable equipment that requires modest amounts of power **batteries** are suitable. For larger powers or in fixed locations where the mains supply is available it is more economic to use mains driven **power packs**. For remote locations such as space satellites it is necessary to use **solar cells**.

In mains operated equipment the mains voltage is first converted into unidirectional pulses, a process known as **rectification**. This is achieved by first transforming the voltage to a suitable value and then using diodes as rectifiers. The usual practice today is to choose **silicon diodes** as the rectifying agency. Depending upon the number and configuration used, these diodes may be connected as **half-wave**, **full-wave** or **bridge rectifiers**.

Filters are needed to smooth out the unidirectional pulses into a steady voltage. Large-value electrolytic capacitors are used in conjunction with smoothing chokes (inductors), or more usually resistors, where the currents involved are not too large.

Where the simple type of power supply is incapable of supplying current at a steady enough voltage, a **voltage stabilising circuit** is required. The simplest circuit is based on the special properties of a zener diode. For more sophisticated control it is necessary to use a 'ballast' transistor which is controlled by an amplifier. This amplifier compares the actual output voltage with a reference voltage (provided by a zener diode) and alters the operating conditions of the 'ballast' transistor in such a way that any deviation of output voltage from the desired value is reduced to negligible proportions. Stabilizing circuits are now readily available in integrated circuit form.

Thyristors are used to control the power to a load in a variable way. If full-wave control is required a **bridge rectifier** is needed. Alternatively, a **triac** (which is an integrated circuit form of two interconnected s.c.r.s) can be used. A.C. power can be controlled in a wide variety of loads such as incandescent lamps, furnaces, heaters and electric motors.

Two types of control are possible, the **phase-shift** method and the **burst-fire** controller. In the first method the supply of power is delayed so that current is not supplied throughout a complete half-cycle. *RC* circuits are used for this. Phase-shift control gives rise to R.F. interference which must be suppressed by using suitable filter circuits. When the powers involved are too large to make this practical the burst-fire method can be used. Here a special circuit controls the number of complete half-cycles during which power is supplied. The method is not suitable for light dimmers or motor controllers.

Questions

1. What are the factors that determine the choice of batteries or mains as a source of power for electronic equipment?
2. Why are metal rectifiers not popular nowadays?
3. Which is more suitable for power control purposes: a thyristor or a triac?
4. Why is the burst-fire method of control unsuitable for use with lamp dimmers?

Suggested Further Reading

Olsen, G. H., *A Course Book for Students*, Butterworth, 1973.

8

OSCILLATORS

An oscillator is an instrument for producing voltages that vary in a regular fashion; the waveforms of the voltages are repeated exactly in equal successive intervals of time. In many cases the waveform of the output voltage is sinusoidal and the oscillator is then called a **sine-wave generator** or **harmonic oscillator**. Those instruments that produce repetitive waveforms that are square, triangular or sawtooth in shape are called **relaxation oscillators**. The term 'relaxation' is used because during the generation of the waveform there is a period of activity in which there is a sharp transition from one state to another. This period is then followed by a relatively quiescent one, after which the cycle is repeated. Several examples of relaxation oscillators occur in nature, the most common one being the heart. During the operation of the heart there is a period of activity in which the blood is pumped through the heart chambers and out into the arteries. This period is followed by one in which the heart muscles relax and prepare for the next burst of activity.

Oscillators can be constructed so as to operate at frequencies as low as one or two cycles an hour or as high as hundreds of megahertz. The selection of a suitable frequency or range of frequencies depends upon the function that the oscillator is required to perform. For the testing of equipment, or as a source of power for energizing a.c. bridges, the frequencies are usually in the L.F. or low R.F. range. Conductivity cells and electrolytic tanks are supplied with energy at frequencies of a few hundred hertz. Radio-frequency oscillators are widely used in the generation of carrier waves for telecommunication systems and in the construction of non-lethal e.h.t. supplies. Industrial heaters of dielectric materials such as wood, glue and plastics depend upon R.F. oscillators. Physiotherapy departments in hospitals use this type of heater in the treatment of bone and tissue disorders. Where the heating of electrically conducting material, such as metal ingots, is involved, induction coils are fed from power oscillators operating at lower frequencies. The material to be heated is placed within the coil and the eddy currents that are induced within the material cause rises in temperature. Both induction heating and dielectric heating have the advantage that the heating is produced within the bulk of the material. These methods of heating do not therefore rely on conduction from a hot surface layer.

For the applications described above the waveform produced by the oscillator has usually to be sinusoidal or nearly so. In other applications such as cathode-ray oscilloscopes, television receivers, radar equipment and digital computers we find it necessary to use relaxation oscillators.

In both types of oscillator (the sine-wave and relaxation types) it is our aim to produce a rapid to-and-fro motion of electrons in a conductor.

152

Sine-Wave Generators (Harmonic Oscillators)

Before considering the generation of sine-wave electrical oscillations let us consider a simple example from the mechanical world. By studying an oscillating mechanical system we may more easily see what is involved in the analogous electronic system.

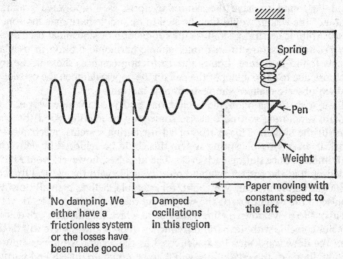

No damping. We either have a frictionless system or the losses have been made good

Damped oscillations in this region

Paper moving with constant speed to the left

Fig. 91. An example of mechanical oscillations.

Fig. 91 shows a weight suspended on a spring. One end of the spring is firmly attached to a motionless platform, the other end being attached to the weight. When first attached, the weight stretches the spring, so the spring, being extended, exerts a force on the weight. It turns out that the force is proportional to the extension of the spring. Eventually the spring is extended to a degree that enables it to exert a force equal and opposite to the force of gravity on the weight. The system then becomes motionless; we say that it is in equilibrium. If we now pull the weight down the force exerted by the spring increases because the extension of the spring has been increased. The force exerted by the spring is now greater than that exerted by the gravitational force so, on releasing the weight, the latter will move upwards. When the weight reaches the equilibrium position the net force on the weight is again zero, but because of its inertia the weight continues to move upwards. At this point some of the potential energy previously stored in the spring is transferred to the weight as energy of motion (kinetic energy). The extension of the spring is reduced and hence the upward tension is reduced. The downward gravitational force is now the greater. The weight gradually comes to rest and then starts to descend. It passes the equilibrium position and eventually reaches its lowest point, whereupon the whole cycle of operations is repeated. The movement from the lowest point to the highest point and back again to the lowest point represents one complete cycle of movement. The time taken to complete one cycle is the **periodic time (or period)** of the oscillating system. A record of the oscillations can be made by attaching a pen to the weight and marking a strip of paper which travels with uniform speed in a horizontal direction from right to left past the weight. If there were absolutely no losses in the system the oscillations would continue indefinitely. The trace on the paper would be found to have the same waveform as a sine-wave graph; the motion of the weight would then be said to be simple harmonic motion. The term 'harmonic generator' implies the production of sinusoidal oscillations.

In practice it is rare to encounter simple harmonic motion in oscillating systems found in nature. Losses due to friction (such as those in the spring and those due to the contact of the pen on the paper) dampen the oscillations. We then observe **damped** simple harmonic motion.

If we wish to maintain the oscillations and produce sine waves with a constant amplitude we must devise some way of making good the energy losses of the system. This is true of all oscillating systems, mechanical and electrical. In our simple spring system this could be achieved by delivering a small impulse once during each cycle. The impulses, however, would have to be delivered at the correct rate and must coincide with the natural frequency of the oscillating system. We could, for example, deliver a small downward impulse to the weight just as the weight had reached the top of its travel and was about to move down again. If the rate at which the impulses are delivered does not coincide with the natural frequency of oscillations there will be times when the downward impulse is delivered as the weight is moving upwards. The amplitude of the oscillations will then be rapidly reduced, and harmonic (i.e. sinusoidal) oscillation will cease.

The natural frequency of oscillation is determined by the size of the weight and the stiffness of the spring. For a given spring, the heavier the weight the slower will be the oscillations. The stiffer the spring, the swifter will be the oscillations. The opposite of stiffness is known as compliance; so we see that the greater the compliance the slower will be the oscillations. Increases in the size of the weight and increases in compliance both reduce the frequency of the oscillations.

The Tank Circuit

Fig. 92 shows a capacitor connected to an inductor. If we assume that initially the capacitor is charged and the switch is open then an electric field exists across the plates of the capacitor; there is a voltage across the capacitor terminals. No current exists in the inductor and hence there is no magnetic field associated with the inductor. All of the energy is stored in the capacitor. (This system is analogous to holding the weight down at a lower position than the equilibrium position. All of the energy is stored in the spring; since the weight is motionless it has no kinetic energy. The compliance of the spring is analogous, or equivalent, to the capacitance and the size of the weight is analogous to the inductance.)

On closing the switch (analogous to releasing the weight), the capacitor starts to discharge and a current is established around the circuit. The magnetic field builds up to a maximum value at which point all of the electric energy is converted into magnetic energy. (In the mechanical system all of the potential energy in the spring is converted to kinetic energy. The inertia of the weight, however, keeps the motion going.) The inductor—which opposes changes of current—keeps the current flowing. This occurs because the magnetic field collapses, inducing an e.m.f. in the coils, and thus charges the capacitor to the opposite polarity. Once the capacitor is fully charged a

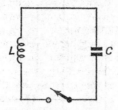

Fig. 92. The tank circuit.

similar discharging action occurs in the opposite direction until we arrive at our initial position. The movement of the electrons to and fro has passed through a complete cycle. If there were no losses in the system the oscillations of the electrons in the circuit would continue indefinitely. The natural frequency of oscillation is determined by the magnitudes of the inductance and capacitance. The natural or resonant frequency is given by the formula $f = 1/(2\pi\sqrt{LC})$.

In practice, losses in the system prevent continuous sinusoidal oscillations. The main loss is due to the resistance of the inductor, which absorbs electrical energy, converting it into heat just as friction converts energy of motion into heat energy in a mechanical system. If the frequency of oscillation is high enough, and the geometry of the coil and inductor are correct, considerable energy can be lost as electromagnetic radiation. These radiations, i.e. radio waves, are very useful when we wish to convey information in the form of radio and television programmes from one location to another. We shall need to refer to this mechanism when we discuss radio transmissions in Chapters 9 and 11. The losses cause the oscillations of our simple system to die away, and the waveform of the oscillating current would be a damped sine wave.

If we wish to maintain the oscillations so as to produce sine waves with a constant amplitude, we must devise some way of making good the energy losses. This we do by delivering a small amount of energy during each cycle in just the right way to make up for the losses. The main *LC* (resonant) circuit acts as a 'tank' of energy which we have to 'top up' at regular intervals to make up for the 'evaporation'.

The *LC* Oscillator

The most general method of making good the losses in a tank circuit, and thus produce sinusoidal oscillations, is to use a feedback amplifier in which the feedback is positive. The large majority of electronic oscillators consist of positive feedback amplifiers which, in effect, produce their own input signal. The frequency of operation is controlled by the nature of the feedback circuit.

Fig. 93(a) shows how this is done in the case of a simple high-frequency oscillator. Here we have a standard amplifying circuit with a biasing potential divider and bypassed emitter resistor. The collector load, however, instead of

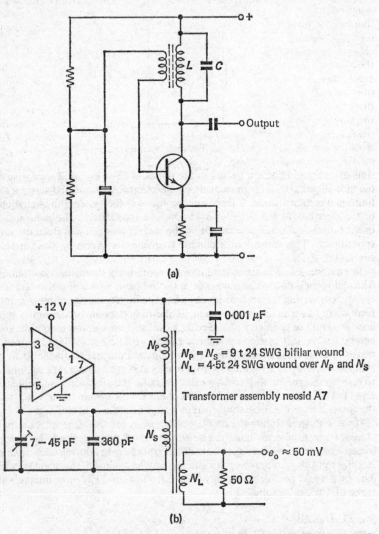

(a)

(b)

Fig. 93. Two forms of *LC* oscillator: (a) tuned collector oscillator; (b) 10 MHz oscillator using an SGS L103T2 R.F./I.F. amplifier.

being a resistor, consists of a tuned *LC* tank circuit. The values of *L* and *C* are chosen so that the frequency of oscillations has the required value. To obtain a good sinusoidal waveform the *Q* value of the tuned circuit must be as high as possible. (The *Q*, or quality, value of the tuned circuit depends almost entirely on the quality of the coil, which must have as low a resistance as possible for a given inductance. The formula for the *Q* of a coil is given by $Q = 2\pi fL/R$.) When the *Q* is high—say, 400 or so—the oscillating currents in the tank circuit are very much greater than the supply, i.e. 'topping-up', currents. Feedback is obtained in this case by coupling the input base circuit to the tank circuit using transformer action. The feedback fraction is dictated by the mutual coupling between the primary coil of the tank circuit and the secondary coil in the base circuit. To avoid too much disturbance of the tank circuit the mutual coupling should be as small as possible consistent with maintaining oscillations. We will then preserve a good sinusoidal waveform. To assist in keeping the coupling loose, a transistor having a high value of h_{fe} at the operating frequency is necessary. Positive feedback is obtained by ensuring that the secondary coil leads are connected into the base circuit in the correct way. The output may be taken from the collector via a capacitor. Alternatively, a second secondary coil coupled to the tank circuit may be used.

The design of transistor oscillators is complicated by the fact that loading of the tank circuit, i.e. taking too much current from the tank, adversely affects both the frequency stability and the waveform. This is serious in communication systems, where there are stringent requirements to keep the frequency of oscillation constant and the distortion down to a very low level. A common solution is to provide some sort of impedance isolation between the oscillating circuit and the load. Fortunately, nowadays the circuit designer's worries are reduced since he can use an integrated-circuit amplifier that has been manufactured for use at radio frequencies.

An oscillator based on the SGS L103T2 R.F. amplifier is shown in Fig. 93(b). Here the tank circuit is connected to the input terminals. Since the input impedance of these amplifiers is so high, the loading on the tank circuit is practically negligible. The output impedance of the L103T2 is low and hence it is possible to drive loads of only a few tens of ohms. If such a low load were coupled via transformer action directly into the tank circuit, the damping would be so great that even if oscillations were maintained, the frequency stability and waveform would be very poor.

The Hartley and Colpitts Oscillators

Two popular and well tried oscillator circuits, that are basically the same, are due to Hartley and Colpitts. From Fig. 94 it will be possible to recognize the two essential ingredients of an *LC* oscillator—namely, the tank circuit and the feedback arrangement. In Fig. 94(a), showing the Hartley oscillator, the tank circuit consists of a parallel combination of C_2 with L_1L_3. Actually L_1L_3 is a single coil provided with a suitable tapping point so as to enable positive feedback to take place. Fig. 94(b), showing the Colpitts oscillator, is basically the same, but the tapping point for feedback is fixed in the capacitor branch.

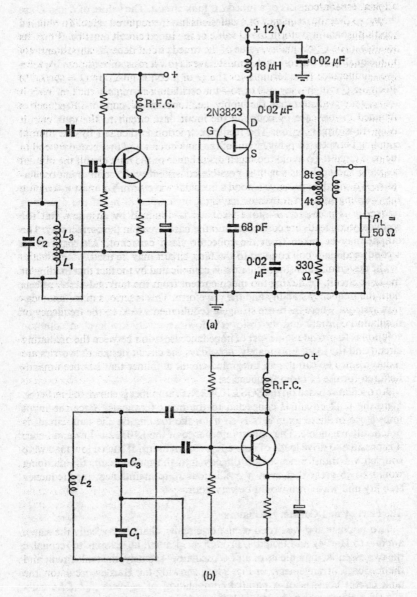

(a)

(b)

Fig. 94. High-frequency oscillators based on the circuits of Hartley and Colpitts:
(a) two versions of the Hartley oscillator; (b) the Colpitts oscillator.

Crystal Oscillators

We have already mentioned the problem of frequency stability with *LC* oscillating circuits. Various factors cause the frequency to drift from the required value. The most important of these is the change in capacitance and inductance that occurs when the temperature changes. Oscillation at a frequency different from the resonant frequency of the tank circuit is caused by the presence of components associated with the tank circuit, such as leads to transistors, interelectrode and wiring capacitances. Changes in the properties of these components, with say temperature or humidity, all contribute to frequency drift. As previously stated, the problems can be largely overcome by using high-*Q* coils and good-quality capacitors with stable characteristics. With ordinary inductors and capacitors, however, *Q*-values in excess of a few hundred are very difficult or impossible to achieve; very large improvements in frequency stability can, however, be achieved by using a quartz crystal in place of the conventional tuned circuit.

It has been discovered that certain crystals (notably quartz and Rochelle salt) exhibit **piezo-electric** properties, that is to say when mechanical forces are applied to the crystal, small voltages are produced across the faces of the crystal. This property is used to good effect in pick-ups for sound reproduction from gramophone records. Here the small mechanical forces produced when the stylus tracks the groove give rise to corresponding voltages at the pick-up terminals. The piezo-electric effect includes the reverse action, i.e. when voltages are applied to the faces of the crystal mechanical forces are developed within the crystal.

Quartz is chosen for oscillator frequency standards because this material is almost perfectly elastic; if mechanical oscillations are started in the crystal it takes a long time for the oscillations to die away. (Compare this with a spring-weight system in which the energy losses are extremely small.) Quartz crystals, therefore, have a very high mechanical *Q*. So far as the electrical properties are concerned, a quartz crystal is equivalent to the *LC* resonant circuit shown in Fig. 95. The values of *L*, *R*, C_1 and C_2 depend upon the physical size of the crystal and how the crystal is cut from the original mass. The crystal itself has conducting electrodes spattered in a vacuum chamber on to two of the faces. Connecting leads are then joined to the spattered electrodes. When the leads are connected to a source of oscillating voltage, mechanical vibrations are set up within the crystal. Provided the frequency of the oscillating voltage is close to a resonant frequency of the crystal plate then the crystal forces the oscillating voltage to assume a resonant frequency determined by the plate, i.e. the frequency of the oscillating voltage is 'pulled in' so as to coincide with the oscillating frequency of the plate. By using the crystal in place of an *LC* resonant circuit in an oscillator the frequency of operation is determined almost entirely by the crystal. *Q* values in excess of 20,000 are easily obtained with readily available crystals.

The actual oscillator circuits follow much the same line as those for a conventional *LC* oscillator. Figs. 96 and 97 give two practical examples. In Fig. 96 the amplifier and feedback arrangements are obvious. Fig. 97 is a crystal equivalent of a Colpitts oscillator.

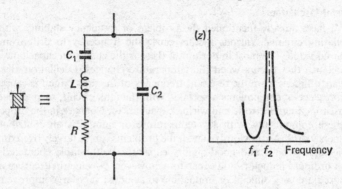

Fig. 95. The circuit symbol for a crystal together with its equivalent circuit. The graph shows the variation of impedance with frequency in the region of series (f_1) and parallel (f_2) resonant frequencies. The values of L, R, C_1 and C_2 depend upon the individual crystal. One typical sample has values of $L = 5.2$ H, $R = 280$ Ω, $C_1 = 0.01$ pF and $C_2 = 6$ pF. The series resonant frequency is 698 kHz and the Q value 81,400.

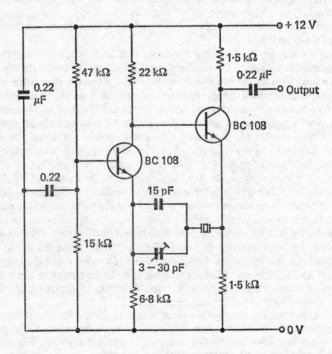

Fig. 96. A crystal oscillator.

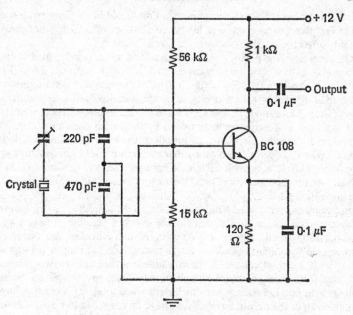

Fig. 97. A crystal version of a Colpitts oscillator.

RC Oscillators

When oscillators are required to operate at frequencies below about 50 kHz, and especially at audio frequencies, it is inconvenient to use *LC* circuits. To obtain the necessary inductance, the size of the coil is inconveniently large, with the result that it is difficult to construct coils with a sufficiently high *Q* value; the amount of wire required is large and consequently the resistance is inevitably high. The sheer bulk of large coils makes them inconvenient for use in transistorized equipment, and such coils are prone to pick up 50 Hz hum signals. The construction of large capacitors necessary for low-frequency operation also presents us with problems. For these reasons oscillators for use at low frequencies are based on resistor-capacitor networks.

The principle of operation is still one of using a feedback amplifier in which the feedback is positive instead of negative. The feedback line consists of suitable frequency-selective networks of resistors and capacitors. It will be recalled that the formula for the gain of a feedback amplifier is given by

$$A' = \frac{A}{1 - \beta A}$$

For negative feedback we arrange that the open-loop gain, *A*, has a 180° phase reversal, so in our formula $A' = -A/(1 + \beta A)$. For practical values of β and *A* we find that the closed-loop gain is usually much less than the open-loop gain. This brings with it all the advantages discussed in the chapter on amplifiers. For this purpose there is no phase shift introduced by the feedback circuit; often the feedback circuit is a simple resistor.

Consider now the position when the feedback circuit is designed to introduce a shift of 180°. β is then negative, so $A' = -A/(1 - \beta A)$. If $\beta A = 1$ (e.g. if $\beta = \frac{1}{1000}$ and $A = 1,000$), we are in trouble because A' then is equal to the value of A divided by zero which is, of course, infinitely large. In practice, the closed loop gain cannot be infinitely large so the amplifier becomes unstable and oscillates at a frequency determined by the circuit constants. If we were intending to use the circuit as an amplifier these uncontrolled oscillations would upset the operation so much that the circuit could not function as an amplifier. This is why with certain integrated-circuit amplifiers (which are always used as feedback amplifiers) we must take special precautions to avoid the condition $\beta A = 1$.

In integrated circuits the trouble often lies in the amplifier itself. Because of stray capacitances which are unavoidable it is not possible to have a phase shift of 180° at all frequencies. At very high frequencies the phase shift becomes 360°, which in practical terms means that the output voltage is in phase with the input voltage. If there is sufficient gain in the system then any electrical disturbance within the amplifier is fed back to the input so the input voltage changes. Let us say that this input voltage rises. This input voltage is amplified so at the output terminals we have an amplified in-phase version of the input. When a fraction of the output voltage is fed back to the input, this feedback voltage will reinforce the original disturbance. This is further amplified and fed back so further reinforcement takes place, and so on until the amplifier saturates. A reverse action then takes place until the amplifier is cut off. This brings about an oscillation between the two limiting states. It will be realized that to sustain oscillations it is necessary to have the condition that any attenuation (i.e. weakening) of the signal in the feedback line is compensated for by sufficient gain in the amplifier—that is to say, βA must equal one.

Although this instability must not be allowed to develop in an amplifier, we can make use of the oscillating mechanism when we wish to build an oscillator. This is why we refer to all oscillators as amplifiers that produce their own input signals by means of feedback. In designing oscillators we need to control the feedback in such a way that $\beta A = 1$ holds at only one frequency—the frequency we require. This can be achieved in an LC oscillator by using a resonant circuit.

Fig. 98 shows how oscillations can be produced with an RC circuit. Here we have a conventional single-transistor arrangement. As it is a single-stage amplifier, the output is 180° out of phase with the input. If we now use three capacitors and three resistors arranged as shown, this RC network will produce a phase shift of 180° when the signal at the input terminals to the network has a unique frequency. It turns out that the frequency is given by $f = 1/(2\pi\sqrt{6}CR)$. At this frequency the output voltage is only $\frac{1}{29}$ of the input voltage. If we now arrange that the gain of the amplifier is 29, the attenuation in the feedback circuit is made up for by the gain in the amplifier. Oscillations are then sustained. In practice, the last resistor in the RC network also acts as

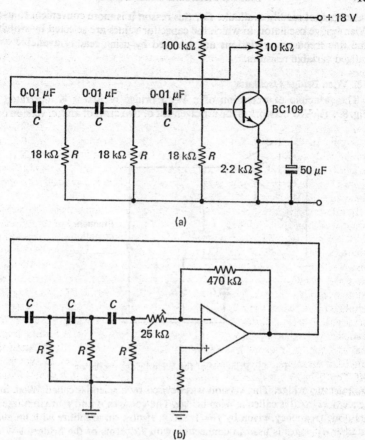

(a)

(b)

Fig. 98. Three-section phase-shift oscillators. (a) If the resistors in the phase-shifting network are altered, care must be taken to preserve the correct biasing of the transistor. The transistor input and bias impedances modify the frequency of operation which departs from $1/2\pi(\sqrt{6})CR$. The potential divider in the collector circuit allows the voltage fed back to be adjusted for the best waveform. (b) An i.c. arrangement that allows an improved calculation of the frequency to be made from $1/2\pi$-$(\sqrt{6})CR$, since the input impedance of the amplifier is large. (The power supply leads are not shown.) The preset resistor should be adjusted for the best waveform.

part of the bias circuit. The bias chain therefore has a small effect on the frequency of operation, which is not quite that obtained from the formula above.

Where the frequency of the oscillator voltage must be variable, simultaneous adjustment of all the resistors or capacitors in a phase-shift network is not convenient because three-ganged variable resistors or capacitors of suitable value are not readily available. For this reason it is more convenient to use a Wien bridge oscillator, in which the capacitor values are selected by switches and fine frequency variations are achieved by using readily available two-ganged variable resistors.

The Wien Bridge Oscillator

The principle of operation of a Wien bridge oscillator is illustrated in Fig. 99. The *RC* network is the arrangement of two arms of an a.c. version of a

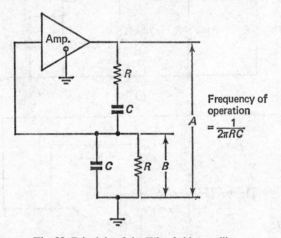

Fig. 99. Principle of the Wien bridge oscillator.

Wheatstone bridge. This version was devised by a scientist called Wien, and hence the circuit is called a Wien bridge. The voltages *A* and *B* are in phase at only one frequency, given by $f = 1/2\pi CR$. If now an amplifier with an even number of stages is used in connection with *RC* arms of the bridge, a Wien bridge oscillator has been made. The amplifier must have an even number of stages so that there is no phase shift between the input and output voltages. The *RC* network forms an a.c. voltage divider and hence a fraction of the output is fed back to the input of the amplifier. By using an even number of stages, the output voltage is in phase with the input voltage. Connecting the bridge to the input of the amplifier, as shown, produces positive feedback at one frequency and oscillations at that frequency are sustained when the gain exceeds three.

Fig. 100 shows a practical Wien bridge oscillator with variable frequency output, designed by Ferranti Ltd. The first stage consists of a Darlington pair. The current gain and input impedance are high, which is desirable when operating this type of oscillator. The high input impedance ensures that the lower half of the Wien bridge is not upset by being loaded. To ensure that the voltage amplification is independent of frequency, negative feedback is introduced into the amplifier. A high initial gain is needed before the application of the feedback to make up for the fall that occurs when negative feedback is

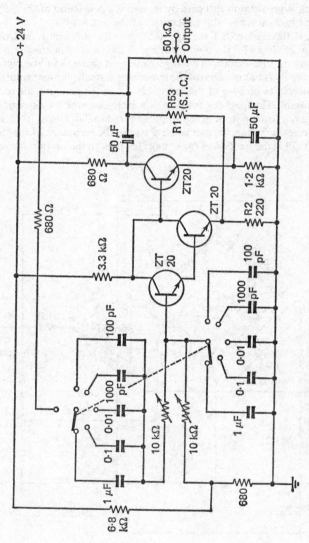

Fig. 100. A practical Wien bridge oscillator to operate over the range 15 Hz to 2 MHz (Ferranti Ltd.).

applied. A further advantage of the use of negative feedback is the low output impedance obtained; the loading effect of the Wien bridge is then minimized. The feedback is provided in this case by R_1 and R_2. A fraction of the output voltage is fed back in series with the emitter of the first stage.

By using a thermistor (S.T.C. type R53) for R_1, amplitude control is achieved. Any tendency for the output voltage to change is counteracted in the following way. A rise in output voltage causes an increase in the current through R_1 and R_2. The resistance of the thermistor, a small glass-encapsulated bead type, decreases because of the inevitable rise in temperature due to the increased current. The feedback fraction thus increases, which automatically reduces the gain; hence output of the amplifier is reduced to almost its former value. Decreases in voltage output cause a rise in the resistance of the thermistor and a fall in the fraction of the voltage fed back to the emitter. A rise in

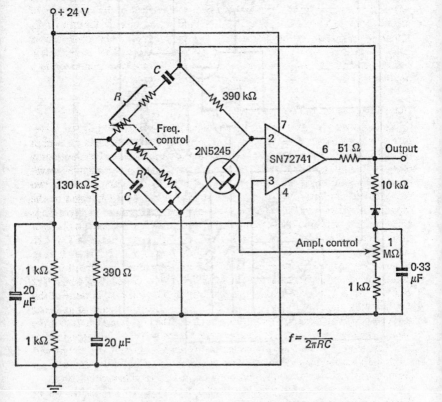

Fig. 101. A Wien bridge oscillator that uses an f.e.t. and integrated circuit. The principles of operation follow Fig. 100. The Texas f.e.t. 2N5245 is used to control the amplitude of oscillation; in effect the resistance of one of the bridge arms is voltage-controlled via the gate.

gain results, which returns the amplitude of the output voltage almost to its former value.

With the capacitor values given, the frequency ranges covered are 15–200 Hz, 150 Hz–2 kHz, 1·5–20 kHz, 15–200 kHz, 150 kHz–2 MHz. The output voltage is about 1 V (r.m.s.). For a change in supply voltage of 4 V, the change in output voltage is less than 1 per cent and the change in frequency less than 2 per cent.

Fig. 101 shows a version using an f.e.t. and an integrated circuit. The *RC* network could follow the lines of Fig. 100.

Relaxation Oscillators

Relaxation oscillators produce repetitive waveforms that are not sinusoidal. Perhaps the two most important examples of these waveforms from a practical point of view are the **rectangular** (so-called **square**) **waves** and the **sawtooth waves**. The square waves are of immense importance in the operation of digital computers and digital apparatus generally. On the entertainment side, square-wave oscillators are essential to the production of electronic music. Electronic organs are based on square-wave oscillators. This is because a square-wave is rich in harmonics. The sawtooth waveform generator is a necessary part of cathode-ray oscilloscopes and television sets.

Generation of Rectangular Waveforms

When we wish to produce oscillating voltages that have rectangular waveforms, it is usual to resort to two-state circuits in which there is an abrupt transition from one state to the other. The **multivibrator** is the most commonly used circuit, and gets its name from the fact that square or rectangular waves are rich in harmonics. The basic circuit is given in Fig. 102, in which two *RC*-coupled amplifiers are used, the output of one being connected to the input of the other, and vice versa. If we regard the arrangement as a two-stage amplifier (Fig. 102(b)) we see that positive feedback occurs since the phase of the collector voltage of Tr2 is the same as that of the base voltage of Tr1. The coupling back of the output voltage of Tr2 to the input of Tr1 is such as to enhance any original disturbance that initiates the action. The circuit is thus rapidly driven into the condition whereby Tr1 is fully conducting whilst Tr2 is cut off. Such a condition is not permanently stable, however, because of the a.c. coupling via the capacitors; the condition is often referred to as being **quasi-stable**. The regenerative switching nature of the circuit is such as to drive the arrangement into its other quasi-stable, whereby Tr2 is fully conducting and Tr1 is cut off. As long as the supply voltage is present the circuit will continually oscillate between the two quasi-stable states; since there is no permanently stable state such a circuit is termed an **astable multivibrator**.

To assist in understanding the mode of operation of this circuit, the waveforms at different points in the circuit are shown in Fig. 102 along with the circuit diagrams. Before considering the action in detail, it will be helpful to recall the way in which a capacitor transfers a signal from one part of the circuit to another. Provided the charge held by a capacitor remains constant,

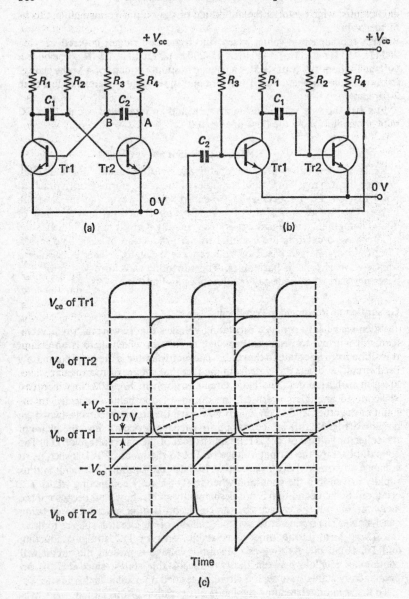

Fig. 102. Astable multivator: (a) basic cross-coupled astable multivator; (b) circuit of (a) redrawn to emphasize the amplifier arrangement with positive feedback.

the potential difference across the plates must also remain constant. If, therefore, one plate (A) of a capacitor is connected to a line having a voltage of $+10$ V and the other plate (B) is connected to a line having a voltage of 0 V, then, after charging, the voltage across the capacitor is 10 V. If now the potential of plate A is suddenly reduced to $+1$ V, the voltage on plate B must fall to -9 V provided there is no change of charge in the capacitor. If charging or discharging of the capacitor can take place via resistors then the voltage changes from -9 V to some new value; the time it takes to change to the new value depends upon the values of resistance, capacitance and supply voltage.

After energizing the circuit, oscillations are initiated by a switching transient or some other circuit disturbance, and very quickly steady-state oscillations are established. At one point in the cycle the base of Tr1 is driven well beyond cut-off. At this time Tr2 is heavily conducting, the collector current is high and the corresponding collector voltage is only a few tenths of a volt above zero. Although Tr1 is sharply cut off, the collector voltage does not rise immediately to the supply voltage because C_1 is charging via R_1 and the low resistance base-to-emitter path in the heavily conducting transistor Tr2. In a short time, however, determined mainly by the time constant C_1R_1, the charging is virtually complete and the voltage of Tr1 then assumes the supply voltage V_{cc}. Meanwhile C_2, which had previously been charged by a similar mechanism to that which charged C_1, is discharging via R_3 and the low resistance path offered by Tr2. Since R_3 is large, the time constant C_2R_3 is long compared with C_1R_1. The base voltage of Tr1 therefore rises rather slowly towards the cut-off potential, during which time the circuit is in a quasi-stable state, the collector of Tr1 remaining at V_{cc} and the collector of Tr2 remaining at almost zero volts. Once the voltage at the base of Tr1 reaches about 0·7 V, Tr1 starts to conduct. (We are assuming the use of silicon transistors.) The collector current in Tr1 rises and the resulting fall in collector voltage is transferred via C_1 to the base of Tr2. The current in Tr2 then falls and the rise in voltage at its collector is transferred to the base of Tr1. A cumulative regeneration action occurs and very rapidly Tr1 is turned on and Tr2 cut off. The collector voltage of Tr1 falls very rapidly from V_{cc} to about 0·15 V and consequently the base voltage at Tr2 is driven to almost $-V_{cc}$. Once the potential at the base of Tr2 reaches the cut-off point, Tr2 starts to conduct again and a further cumulative action develops.

The output from this oscillator may be taken from either collector. The mark-space ratio can be altered by selecting appropriately different time constants for C_2R_3 and C_1R_2.

In a Made Simple book it may be appropriate to state that both sine-wave oscillations and square-wave oscillations can be generated very simply from the constructor's point of view by employing the integrated circuit shown in Fig. 103.

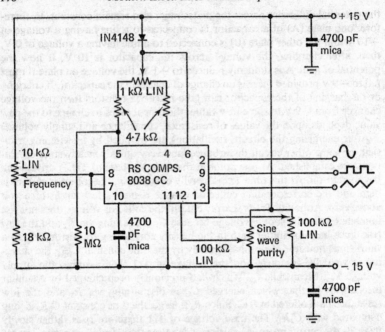

Fig. 103. An audio oscillator that produces three waveforms with frequencies in the range 20 Hz to 20 kHz. The I.C. is an RS Components type 8038CC (order number 305–844). Also available is a more expensive improved version type 8038 BC (order number 306–314).

Summary

An **oscillator** is an instrument for producing voltages that vary with time in a regular fashion. For sinusoidal oscillations one complete oscillation is a **cycle**; **frequency** (f) is the number of cycles completed in one second; **period** is the time taken to complete one cycle and is equal to $1/f$.

Damped oscillations die down because of internal losses in the system. To produce continuous electrical oscillations these losses must be made good by supplying energy from an amplifier.

LC oscillators depend upon a **tank circuit**; the natural frequency of oscillation of this circuit is $1/(2\pi\sqrt{(LC)})$. Many circuits for LC oscillators have been devised; all of them depend upon the controlled feedback, at one frequency, of a fraction of the output of an amplifier from the output terminal to the input terminal. Positive feedback is required for oscillations to be sustained.

The frequency stability of such an arrangement can be greatly improved if the LC circuit is replaced by a suitable crystal.

For low-frequency oscillations positive feedback is achieved by using RC circuits.

Relaxation oscillators produce repetitive waveforms that are not sinusoidal. Positive feedback must still be used, but this type of oscillator usually depends upon two-state, i.e. bistable, circuits. Bistable circuits, as such, produce square waves, but by passing such waves through differentiating or integrating circuits pulses and triangular waveforms respectively can be produced.

Questions

1. Why are oscillators of such great importance in electronics?

2. Is it possible to generate electrical oscillations in any way that does not involve feedback amplifiers?

3. What are the advantages and disadvantages of using piezo-electric crystals in oscillator circuits?

4. Low-frequency oscillations are to be produced using *RC* circuits. Would you choose a Wien bridge circuit or a phase-shift circuit?

9

RADIO TRANSMISSION AND RECEPTION

Oscillators of the type discussed in the previous chapter are to be found in many different kinds of electronic equipment, but by far and away their greatest application is in the field of radio and television communication. Every radio transmitter must have an oscillator, as must nearly every type of radio receiver. In fact, electrical oscillations are central to all communications systems.

Information in coded form, or as speech, music or pictures, cannot be transmitted from one point to another by means of a continuous stream of sine waves. Although these waves are the simplest oscillations possible, some means must be found to alter their characteristics in accordance with the information which is to be transmitted. One of the simplest ways of doing this is to interrupt the train of waves at specified intervals. Thus we may transmit a short train of waves or a long train of waves. If we identify the short train as a dot and the long train as a dash then a code can be devised such that combinations of dots and dashes represent letters of the alphabet and numerals. Readers will be aware of the code devised by **Morse**, which is perhaps the most famous code ever invented. Whilst such a code can be used to transmit information it is very limited in its application. There is no ready recognition by the vast majority of people, and the information cannot be conveyed with any great speed. It is not surprising therefore that scientists worked hard to enable speech, music and pictures to be transmitted.

The transmission of speech and music as **acoustic**, i.e. sound, waves is not very practical. To reduce losses, the sound pressure variations could be confined within a pipe. Indeed, before the invention of the telephone, communication between master and servant in separate rooms of large houses or between the captain of a ship on his bridge and the engineer in the engine room was possible by using hollow flexible pipes. Such an arrangement is very limited. A great increase in range is possible if the sound pressure waves are first converted by a microphone into **electrical signals**. These signals can then be processed by amplifiers, and transmitted along wires. At the receiver end the electrical signals are converted back into sound waves. If we have designed our system with skill, the sound waves at the receiver end are a reasonable reproduction of the original sound waves at the transmitter end. The microphone, amplifier, receiver and wire system must all be capable of handling signals in the audio frequency range. For telephone purposes it is sufficient to limit this range to that of speech frequencies. The confidential nature of telephone conversations makes it desirable to limit, by a wire system, the number of people who can listen in to the information given by the person at the transmitter.

172

For many purposes—for example, when transmitting domestic entertainment programmes—it would be uneconomic to have a wire from the transmitter to every receiver. Under these circumstances the information is transmitted in the form of **electromagnetic radiations**. Anyone who then wishes to receive the information can, by means of an aerial, detect the presence of these electromagnetic radiations. The small currents induced in the aerial by the electromagnetic radiation are then processed electronically by the receiver. It is easy to see how the old-fashioned term 'wireless set' arose. At the time of its invention it was quite surprising to have communication over long distances without the aid of wires.

Electromagnetic radiations are still not well understood. All wave motion seems to need a medium by which the motion can be transmitted. For example, sound waves are transmitted by causing pressure variations to be set up in the air; sound can also be transmitted in water and through solids. Waves can also carry energy over the surface of liquids as, for example, the ripples on a pond. Wave motion can also be observed in strings made from wire or 'cat-gut'. No one, however, knows how electromagnetic waves are 'carried' throughout the universe. By his very nature man tries to explain new phenomena in terms of those facts already known to him. Since it seems that all wave motion known to him has a medium of propagation therefore electromagnetic waves must also have a medium of propagation. All attempts to find such a medium have so far failed. It may turn out that no medium is necessary; our knowledge in many fields is unfortunately woefully deficient. However, the fact that electromagnetic waves exist cannot be denied. Light, heat rays and X-rays are just three manifestations of such radiation; wireless or radio waves a fourth. All four are essentially the same, differing only in wavelength.

The Launching of Radio Waves

Although the production of electromagnetic waves is a complicated process, and sophisticated mathematical techniques are required to describe it, we can obtain a reasonable qualitative explanation by considering Fig. 104. The tank circuit is shown in Fig. 104(a); we recall that if such a circuit suffered no losses then electrical oscillations would persist indefinitely. At one instant all the energy is stored in the capacitor as an electric field; at another time the energy is stored entirely as a magnetic field associated with the inductor. As the current surges first one way and then in the reverse direction, the energy oscillates between the coil and the capacitor. In practice, losses occur mainly as heat dissipated in the resistance of the coil and leads.

Imagine now that the distance between the plates of the capacitor increases, as in Fig. 104(b). As the current passes along the leads to the capacitor, a magnetic field, concentric with the leads, is established. If the capacitor is now 'opened out' until in effect it consists of the two long leads alone, separated by air (Fig. 104(c)), we find that such an arrangement loses considerable energy. This loss cannot be accounted for by the resistive losses in the circuit. We find that the energy has been radiated as an electromagnetic field.

If now one end of the coil is connected to earth and the other to a long horizontal wire (the aerial), then the capacitor consists of the wire and the

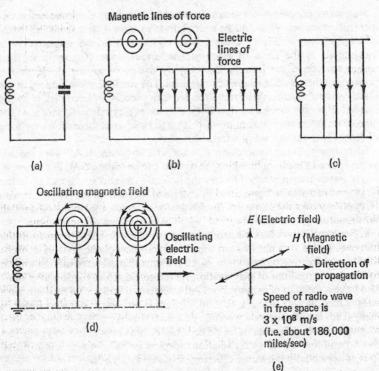

Fig. 104. Generation of an electromagnetic radio wave.

earth—which is a conductor. From Fig. 104(d) we see that the magnetic field is at right angles to the electric field and that the resultant electromagnetic radiation travels at right angles to both these components (Fig. 104(e)). The velocity of propagation is very high—about 300 million metres per second (3×10^8 metres per second) or about 186,000 miles per second. This is also, of course, the velocity of light, which is not surprising since light is also an electromagnetic radiation.

The length of the aerial determines the capacitance which, as we have already seen, is a factor in determining the frequency of oscillation. In practice, we usually specify the frequency and then arrange that the aerial has the correct dimensions. For very high frequency work the aerial needs only to be quite small. A simple dipole may then be used as shown in Fig. 105. This dipole is energized via a feeder line from a tank circuit, which forms part of the transmitter circuit. It is usual to make each pole of the dipole a quarter of a wavelength long. We can easily calculate this length if we know the frequency of operation because

$$\text{Wavelength} \times \text{Frequency} = \text{Velocity of propagation}$$
$$\lambda \qquad\qquad f \qquad\qquad v$$

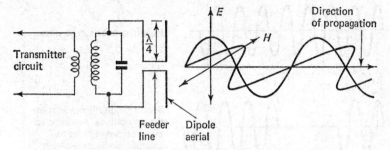

Fig. 105. Launching a radio wave from a dipole aerial.

If, for example, the frequency is 100 MHz, i.e. 10^8 Hz, and we know $v = 3 \times 10^8$ metres per second, then the wavelength is $v \div f$, i.e. 3 metres. Each pole of the dipole therefore needs to be 0·75 m long, i.e. almost 30 inches long.

Consider now the position when we wish to transmit radiations of audio frequencies. Suppose, for example, we choose a frequency of 30 Hz; the wavelength, λ, would then be 186000/30, i.e. 6,200, miles. At 3000 Hz the wavelength is 62 miles. Aerials of comparable dimensions are clearly far too long to be practical; even a quarter-wavelength aerial needs to be more than 1,500 miles if frequencies as low as 30 Hz are to be transmitted. If aerials of practical dimensions are used the frequency of the radiations must be much higher than an audio frequency. Even though we can detect the presence of such waves a direct conversion into an audio signal would be useless since our ears cannot respond to frequencies greater than about 20 kHz. If, therefore, radio waves are to convey information, some feature of the wave must be altered in accordance with the information. When we alter a sine wave in this way we are said to **modulate** it. We may alter the amplitude of the wave in which case we are using **amplitude modulation**. Alternatively, we can keep the amplitude constant and vary the frequency in accordance with the signal information; in this case we are using **frequency modulation**. Electronic equipment, in the form of a suitable radio receiver, is needed to extract the information from the modulated wave. Part of the receiver must therefore be a demodulator, or detector of the high-frequency signals.

Amplitude Modulation and Detection

We have already met one form of amplitude modulation. By stopping and starting a transmitter the amplitude of the radio waves is either zero or some maximum value. This is the basis of sending Morse code messages. For speech and music, however, we need to vary the amplitude between the limits of zero and a maximum in accordance with the programme material.

Fig. 106 shows how the waveform appears when a single audio tone is to be transmitted. The amplitude of the high-frequency wave is varied in time with the waveform of the single tone. Since this tone is being conveyed by the high-frequency radiation such a radio wave is called a **carrier wave**. The degree of modulation is an important consideration since the quality of the sound

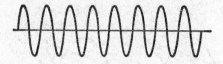

High-frequency carrier wave

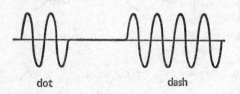

Severe amplitude modulation·
since amplitude is reduced to
zero at times in order to
transmit morse code signals

dot   dash

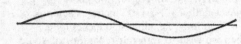

Single audio tone

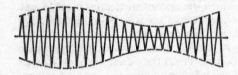

High-frequency carrier wave
amplitude-modulated by audio tone

Fig. 106. Amplitude-modulated waves.

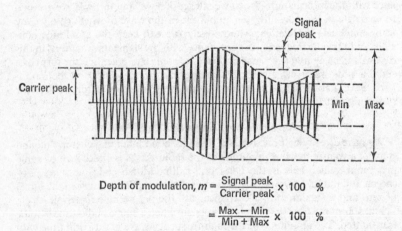

Depth of modulation, $m = \dfrac{\text{Signal peak}}{\text{Carrier peak}} \times 100$ %

$\qquad\quad = \dfrac{\text{Max} - \text{Min}}{\text{Min} + \text{Max}} \times 100$ %

Fig. 107. Definition of modulation index.

eventually obtained after detection depends upon the amount by which the amplitude of the carrier wave is modulated. The amount of modulation also determines the strength of the signal being radiated by the transmitter. The depth of modulation is conveniently expressed as an index or percentage and is defined as shown in Fig. 107.

Sidebands

When we speak of a single frequency in connection with a wave we necessarily imply that the wave must be sinusoidal. If the wave were not sinusoidal then we have a fundamental and other frequencies present. Since an amplitude-modulated wave is not a pure sine wave, frequencies other than that of the carrier must be present. It turns out that if we modulate the amplitude of a carrier wave with a single tone then three frequencies can be shown to be present. These are: (a) the carrier frequency, (b) the carrier frequency plus the tone frequency, and (c) the carrier frequency minus the tone frequency. When speech and music programmes are being transmitted then audio frequencies are involved. If, therefore, a carrier wave is modulated in accordance with the programme material we find that in addition to the carrier frequency we have a band of frequencies extending over the range of the carrier frequency plus all those frequencies present in the audio signal. We have also another band extending over the range of the carrier frequency minus the audio frequencies. These bands are known as **sidebands**. For example, if the carrier frequency is, say, 1 MHz (about in the middle of the medium-wave band) and the audio signal has frequencies in the range of 20 Hz to 10 kHz, then the electromagnetic radiations from the aerial extend over a frequency range from 1 MHz $\pm$ 10 kHz, i.e. 990 kHz to 1·01 MHz. Fig. 108 illustrates the position. To broad-

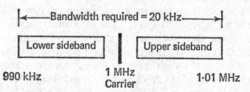

Fig. 108. An amplitude-modulated wave can be analysed into the carrier wave plus sidebands. For audio modulation up to 10 kHz each sideband has a 10 kHz width. The total bandwidth required is therefore 20 kHz. This is the amount of 'space' required in the medium-wave band.

cast audio frequencies up to 10 kHz using a 1 MHz carrier, we therefore need a transmitting channel having a bandwidth of 20 kHz. The medium-wave band extends from about 500 kHz to 1·5 MHz. We therefore see that there is a limited amount of 'space' available if we are to avoid interference between separate stations. It is clearly necessary to use different bands to accommodate the thousands of transmitters that need to operate. The congestion on the broadcast bands has become so severe that we now have recourse to short-range frequency-modulated systems in order to improve the quality of reception and avoid interference from other stations.

Electromagnetic radio waves of the type we are discussing are little affected by the air of the atmosphere, snow and rain, and are able to pass through buildings and other non-conductors with ease. They cannot, however, penetrate conductors such as sheets of metal and wire meshes.

The method by which such waves reach the receiving aerial depends mainly on the wavelength involved. The so-called **ground** or **surface waves** are guided to some extent by the earth and travel just above its surface. Because such waves are affected by the electrical characteristics of the ground over which they are travelling and because the absorption of such waves becomes severe as the frequency increases, ground waves are generally useful at low frequencies. The 'long' waves, having a wavelength in excess of 1,000 metres, can be received reliably over distances of up to 1,000 miles. Much of the ship-to-shore communications is carried out by means of radios using the long-wave band. At the frequencies involved in television and F.M. radio the attenuation (i.e. reduction in strength) of the ground wave is so severe that reception is possible only within an area of a few miles from the transmitter.

The **ground-reflected wave** relies on the fact that the earth can act as a mirror for radio waves. Since such a wave is not subject to continuous absorption by the earth, the range is considerably increased. The reflected-wave may have a very high frequency and so, together with the direct wave (see below), provides the main means for the transmission of F.M. and television signals. A high aerial is necessary to ensure a reasonably large service area.

The **direct**, or **line-of-sight**, **wave** travels directly between the transmitting and receiving aerials. Because of a slight refraction, or bending, in the atmosphere, the 'line-of-sight' is slightly greater than an optical 'line-of-sight' distance. The direct wave becomes of increasing importance as the frequency increases since the radio 'line-of-sight' approaches the optical 'line-of-sight'. Ultra-high-frequency transmissions and those associated with radar, microwave links and air navigational aids all rely on direct waves.

Sky waves make around-the-world short-wave communication possible. Just as the conducting earth makes reflection possible, similar reflections are observed from layers of ionized gases in the earth's upper atmosphere. These ionized layers act as a radio mirror, and are generated by the radiations from the sun at heights of about 50–250 miles above the earth's surface. Several layers have been discovered, the most important being the **Kennelly–Heaviside Layer**, named after the names of the discoverers, and the **Appleton Layer**. The former, found at heights of about 50–100 miles makes possible transmissions above 20 MHz over distances of 1,500 miles. The Appleton Layer, at heights of about 100 to 250 miles, is responsible for reflections at night; the Heaviside Layer is almost absent during the hours of darkness.

Communication by sky waves is unreliable since the 'mirror in the sky' varies in position and density depending upon the sun's activity, seasonal variations and yearly variations. Fig. 109 shows diagrammatically the communication paths taken by sky waves. For frequencies above a critical value the radiations are refracted, pass through the layers and are then lost in outer space. Below a certain critical frequency, waves are refracted within the layers in such a way that total internal reflection can take place. We can compare this with the total internal reflection of light at the interface of two transparent media, provided certain critical conditions obtain. Even if the frequency of the transmissions is below the critical frequency the waves are lost into outer space if the vertical angle at the transmitter is too small. As the vertical angle of the sky wave is increased we eventually reach a critical angle where the waves, instead of passing through the layers, are bent round so much that they

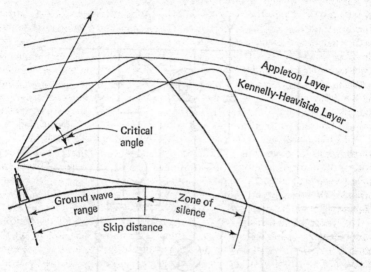

Fig. 109. Reflections of radio waves from the upper atmosphere. The ground wave range depends upon frequency. For very high frequencies that are not reflected by the ionized layers range is only line-of-sight. For long and medium wavebands, waves follow the earth's surface but are eventually absorbed.

are reflected back to earth. The distance between the transmitter and the point on the surface of the earth where the waves are first received after being reflected is known as the **skip distance**. There is a zone within the skip distance where no reception is possible. This zone is just beyond the range of reception of the ground waves and extends to the point where the sky wave is first received. If the sky wave is strong enough it is reflected at the earth's surface and again by the ionized layer. In this way a sky wave may travel around the earth making world-wide reception possible.

The subject of transmitters is a complicated one and not really within the scope of this book. The general principles of operation of a transmitter of amplitude-modulated waves, however, are not difficult to understand. Basically we need an oscillator to generate the carrier wave, and an amplifier to produce audio signals of suitable strength. Amplitude modulation is then carried out by mixing the carrier and audio signals in a special way. The **mixer** is a non-linear device—that is to say, its output is not directly proportional to the input signal. When two signals are fed into a mixer of this type we find that the output consists of the carrier wave whose amplitude varies according to the strength of the audio signal. Fig. 110 shows the circuit for a simple transmitter of the type used in the radio control of model aircraft and boats. (It should be pointed out in passing that the operation of transmitters is not legal in the United Kingdom and many other countries without an appropriate licence.)

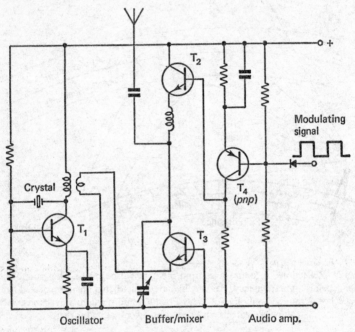

Fig. 110. A simple transmitter.

The various stages of the transmitter will be readily recognized. The oscillator section is of the type already discussed in the previous chapter. Crystal control of the frequency is essential in order to satisfy the stringent G.P.O. regulations in the United Kingdom. The audio signal, after being amplified by a conventional amplifier, is fed to the mixer/buffer section. A *pnp* transistor is used here in order to preserve the correct bias condition for transistor T_2. Transistor T_3, which is transformer-coupled to the oscillator, acts as a buffer, thus preserving the stability and waveform of the oscillator. It will be seen that the load for T_3 is not a pure resistance, but is transistor T_2. Since the effective resistance presented by T_2 depends upon the instantaneous value of the audio signal, it will be seen that the gain of T_3 varies in accordance with the audio signal; the carrier frequency signal has a greater amplitude when the gain is high than it has when the gain is low. We thus see that amplitude modulation has been effected.

Radio Receivers for Amplitude-Modulated (A.M.) Waves

The varying currents within the transmitting aerial give rise to electromagnetic radiations. When these radiations pass through a wire suspended above the earth's surface voltages are induced in the wire. The voltages are very small, being of the order of microvolts, i.e. millionths of a volt, but their waveforms are a replica of those present in the transmitting aerial. We have thus achieved a transfer of information from the transmitter to the receiving location. The tiny signals in the receiving aerial must now be processed electronically so as to recover the information in the form of a reproduced audio signal.

Many signals are induced in the aerial by the simultaneous operation of hundreds of transmitters throughout the world. Our first problem is to isolate the particular signal in which we are interested. This we do by means of a **tuned circuit**. This circuit consists of an inductor connected in parallel with a capacitor, and is in fact the tank circuit discussed in the previous chapter. The tuned circuit exhibits the properties of resonance. In an electronic context this means that the response of the circuit is particularly noticeable when the frequency of the applied signal is the same as the resonant frequency. When the frequency of the applied signal is greater than or less than the resonant frequency the currents induced in the tank circuit are small. The quality of the response depends upon the Q value of the coil ($Q = 2\pi fL/R$). When the Q is high (say 200 in a practical case), the response is excellent at the resonant frequency, and the tuned circuit is able to select one signal from many very effectively. On the other hand, a tuned circuit with a low Q value, say 50, is not able to discriminate very well between carrier waves of roughly similar frequencies.

The position is neatly summarized in what is called a response diagram, an example of which is shown in Fig. 111. It will be seen that if we wish to have a receiver that responds to only one signal, a sharp resonance curve is essential.

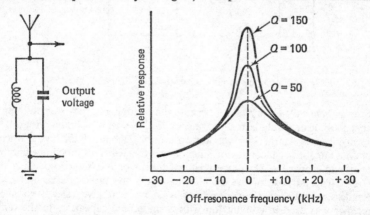

Fig. 111. Relative response showing the effect of the Q-value on the sharpness of the resonance curve.

If the tuned circuit has a Q value of only 50 the response at the resonant frequency is not much greater than that for carrier waves having frequencies 20 kHz greater or less than the resonant frequency. In practical terms, the selectivity of the circuit is poor, and a receiver based on a single tuned circuit would suffer from a good deal of interference from unwanted stations.

Having isolated the signal we wish to receive, the next step is to recover the intelligence (i.e. information) from the modulated carrier wave in the form of an audio signal. This is achieved by the **demodulator** or **detector section**. Although the process of recovering the audio information is correctly called detection, the electronic mechanism is identical with half-wave rectification. This subject has already been discussed in the chapter on power supplies, and so we need not repeat the details here. For radio purposes, of course, the power involved is negligible and the frequency of operation is much higher than that in mains power supplies. The diode is usually a point-contact type because its performance is much better at radio frequencies than that of the *pn* junction power rectifier. The 'smoothing' capacitor is of course much smaller because the frequency of operation is so high.

Fig. 112 shows the simplest possible radio receiver. It is, in fact, a circuit well-known to wireless enthusiasts in the 1920s and 1930s. In those days the

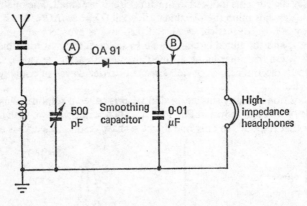

Fig. 112. A simple radio receiver (see also Project 1, page 288).

receiver was called a crystal set because the rectifying agency was an impure crystal of lead sulphide (called galena) together with a brass spring. The spring, or cat's whisker as it was known, was made to touch the crystal whereupon a metal-to-semiconductor junction was formed. This junction exhibited rectifying properties. The modern equivalent is a point-contact diode, e.g. the OA91. The purpose of the diode is to rectify the radio frequency signal. Readers will recall that in converting the a.c. mains to give a steady voltage a rectifier was used in conjunction with a smoothing capacitor. In the radio application we do not require a steady voltage, but need to recover only the audio signal from the modulated wave. The principle is the same, however.

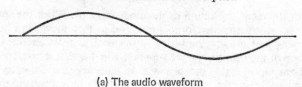

(a) The audio waveform

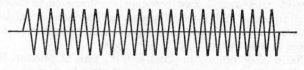

(b) R.F. carrier wave

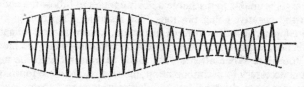

(c) Amplitude-modulated wave

(d) Half-wave rectified version of (c)
with smoothing capacitor absent

(e) Audio signal obtained when smoothing
capacitor is present

Fig. 113. Various stages in the production and detection of amplitude-modulated waves.

Fig. 113 shows in diagram form the various waveforms associated with the broadcasting system. Figs. 113(a), (b) and (c) show the audio wave, R.F. carrier wave and the A.M. wave respectively. It is the waveform at (c) that is present at the point A in the circuit diagram of Fig. 112. At the point B the waveform becomes the half-wave rectified version of the amplitude-modulated wave because of the rectifying action of the diode detector. When the smoothing capacitor is present the waveform is modified to that shown in Fig. 113(e). The charging and discharging action is identical to that described for the rectification of the mains in power supplies. The diagram shows a somewhat 'serrated' waveform, but in practice we must remember that for every cycle of the audio signal the carrier wave goes through about a thousand cycles for medium-wave broadcasts; for television broadcasts on the 405-line system the carrier wave may go through as many as 100,000 cycles for every cycle of the audio waveform. The audio waveform output of the simple receiver is therefore much smoother than Fig. 113(e) suggests. The audio waveform is then passed through a pair of headphones, whereupon the programme material is reproduced as sound waves.

Two forms of distortion can occur with this type of detector. First, the smoothing capacitance must not be too large otherwise the audio waveform cannot be followed. Fig. 114(a) shows the effect. The smoothing capacitor charges on the rising portion of the audio waveform, but, because it holds too much charge, is unable to discharge quickly enough to follow the audio waveform on the negative-going portions. The resulting distortion is known as **diagonal clipping**. Another form of distortion, known as **peak clipping** (Fig. 114(b)), is not the fault of the detector circuit but arises when the modulation index is too high. When attempts are made to transmit with the maximum power it is necessary to use modulation indices that are close to unity. It can be shown that when the modulation index m is equal to 1 then peak clipping occurs. It is necessary in practice to limit m to values no greater than about 0·9.

The simple receiver of Fig. 112 is now mainly of historic interest, although those just starting their studies of electronics may find it instructive and stimulating to receive their first signals from a simple set they have built themselves. Such a receiver suffers from two major disadvantages. First, since only a single tuned circuit is used, the selectivity of such a receiver is not high enough to prevent interference from transmitting stations operating with carrier frequencies close to that of the wanted station. Second, there is no provision of amplification; consequently the receiver is not very sensitive, nor can it operate a loudspeaker directly.

Tuned R.F. Amplifiers

The difficulties inherent in the simple 'crystal' receiver can be largely overcome by using tuned R.F. amplifiers. Such amplifiers are not very different from the audio amplifiers we have already studied except that instead of using a resistor as the collector load, a tuned circuit is used instead. Off resonance, i.e. at operating frequencies above and below the resonant frequency, the impedance of a tuned circuit is very low. As we approach the resonant frequency the impedance rises until at the resonant frequency the impedance becomes very large. (If the resistive losses in the coil and capacitor were reduced to zero then the impedance would become infinitely great. In practice these losses are always present, but are small in a high-Q circuit.)

The amplification of an amplifier stage depends upon the impedance in the

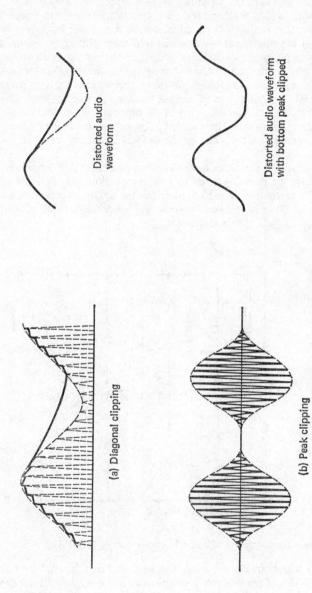

Fig. 114. Distortion of audio output from the detector (a) due to diagonal clipping because of use of too large a smoothing capacitor, and (b) because of over modulation, resulting in clipping of the negative peaks.

collector line. Provided the amplifier stage is operating satisfactorily, the larger the collector load the larger is the amplification or gain of the stage. We see, therefore, that if a high-Q circuit is used as a collector load, very high amplifications are possible at the resonant frequency. Naturally, we arrange that the resonant frequency is variable so that carrier waves of different frequencies can be accommodated. Altering the resonant frequency of the tank circuit is equivalent to tuning in to the station we require. In practice, a ganged variable capacitor is used so that turning one control knob simultaneously alters the resonant frequency of the tuned circuits in both the aerial circuit and the collector line. The resonant frequencies of both circuits must, of course, be the same, i.e. the collector circuit must accurately track the aerial circuit.

Fig. 115 shows the circuit of a **tuned radio-frequency (T.R.F.) receiver**. In order to reduce the receiving aerial to small proportions we often wrap the wire around a **ferrite rod**. The ferrite rod has the ability to concentrate the electromagnetic flux within a small volume and hence we can induce suitable voltages in aerial coils of reasonable size.

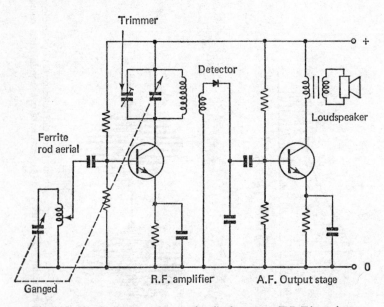

Fig. 115. Circuit diagram of a tuned radio-frequency (T.R.F.) receiver.

Since two tuned circuits are used, the selectivity is improved over that obtained with the simple receiver. The sensitivity is also improved because we have introduced a stage of R.F. amplification before the detector circuit. It may be thought that the sensitivity could be further improved by adding additional R.F. amplifier stages. Although theoretically possible, it turns out that in practice several difficulties arise. First, the noise becomes a problem in sensitive receivers; atmospheric interference in the form of lightning and static, and man-made electrical interference from welders, electric motors, car ignition systems, etc., show large amplitude variations over a wide radio-frequency spectrum. A.M. receivers are particularly affected by these noise sources. Second, the number of stages that can be used is limited because of instability problems. It is practically impossible to build a very high gain R.F. amplifier merely by connecting several stages together. Each stage is obviously operating at the same frequency and it is very easy to have positive feedback from the latter stages to the input. Even though the voltage fed back may be very small, because of the high gain of the system, the conditions for self-oscillation are nearly always present.

A third problem has to do with tracking. Every stage has its own tuned circuit and it is necessary to vary the resonant frequency of each tuned circuit in such a way that all tuned circuits have the same resonant frequency. Apart from the practical difficulty of obtaining ganged capacitors with many sections, the inevitable stray capacitances in the various parts of the circuit upset the matching of each stage.

A fourth problem has to do with fidelity of the final audio output. Although the addition of each tuned circuit improves the selectivity, and excellent sensitivity is obtained, the overall response becomes very sharp. It is as though we were using a single tuned circuit with an extremely high Q value. In order to receive the programme information we must have a receiver which can respond not only to the carrier but *also to the sidebands*. As the receiver becomes more and more selective, increasingly the sidebands are eliminated. The audio frequency response therefore becomes restricted. Ideally we would like the receiver to have a response that look likes a square wave, with the carrier in the centre of the square wave, and the latter to be wide enough to accept all of the sidebands, but thereafter to discriminate severely against stations operating on adjacent carrier frequencies. Such a response is not possible in a highly selective and sensitive T.R.F. receiver.

The Superheterodyne Receiver

Most of the difficulties mentioned above can be eliminated by using the superheterodyne principle. Today all commercial A.M. radios are super-heterodyne receivers. Heterodyne is a term used to describe a method of imposing on a continuous wave another of slightly different frequency so as to produce beats. Readers may be familiar with the sound produced by two tuning forks which, when vibrating, produce tones which differ slightly in frequency. In addition to the two tones a difference or beat tone can also be identified. Whenever the tones from the fork are in phase they reinforce each other and the sound is increased; whenever the two tones are out of phase a certain amount of cancellation takes place and the composite sound is weakened. The rate at which a strengthening or weakening of the sound takes place is determined by the difference in frequency of the two original tones. If this difference frequency lies in the audible range we become aware of a third beat frequency tone.

Heterodyning is essential in Morse Code receivers. In order to obtain reasonable ranges the carrier wave that is started and stopped according to a dot, dash code must have a frequency in the radio range. Since our ears do not respond to radio frequencies some means must be found to enable us to be aware of the presence of the carrier wave. This is achieved in the receiver which, in addition to the tank circuits and R.F. amplifiers, also has an oscillator. The operator adjusts the oscillator frequency so that the difference between the carrier wave frequency and the oscillator frequency lies in the audible range. The heterodyne circuit mixes the oscillator output with the carrier wave, whereupon beats are produced. Since the beats have an audible frequency, they may be amplified in the usual way by an audio amplifier and finally become audible in a pair of earphones or loudspeaker.

This simple system works well when only a single tone is involved to carry the signal information. When speech and music are involved the carrier wave must be modulated in a more complicated way. Nevertheless, a somewhat similar principle can be used in that the incoming carrier wave may be mixed with a local oscillator in the receiver so as to produce difference frequencies. The difference frequencies must not be heard as an audio sound, however, or there would be interference with the wanted programme material. This difficulty is overcome by heterodyning at a supersonic frequency (i.e. a frequency above that of an audible tone); thus we have a supersonic heterodyne receiver. 'Supersonic heterodyne' is a somewhat long term so it is shortened to **superhet(erodyne)**.

The basic parts of a superheterodyne receiver generally consist of a ferrite rod aerial to pick up the transmission signals, an oscillator and mixer stage which is so designed that the difference between the incoming carrier frequency and the oscillator frequency (known as the intermediate frequency) is always the same, one or two stages of amplification of the intermediate frequency (I.F.) signal, a detector to recover the audio information, and an audio amplifier to drive the loudspeaker.

Fig. 116 shows the system in block diagram form. How does such a system overcome the difficulties inherent in a T.R.F. receiver? The superheterodyne receiver converts all incoming carrier frequencies to a fixed intermediate frequency and thus, compared with a T.R.F. receiver, the superheterodyne exhibits much increased sensitivity, stability and selectivity. The increased sensitivity arises because we can employ several I.F. amplifiers (although two is generally all that is needed). Since the frequency is fixed we have no tracking problems as each tuned circuit obviously needs to be tuned to one frequency; no variable capacitor is required. The stability problems are minimized because feedback at the lower intermediate frequency is much less of a problem than at the higher carrier frequency. Selectivity is improved because we use several tuned circuits. However, the superheterodyne does not suffer from the degradation of audio quality inherent in a T.R.F. receiver because a bandpass response close to the ideal is possible. This is achieved by using I.F. transformers. By varying the coupling between the primary and secondary coils we can obtain an excellent response in that part of the frequency range which includes the carrier and the sidebands, whilst at the same time discriminating sharply against adjacent carrier signals.

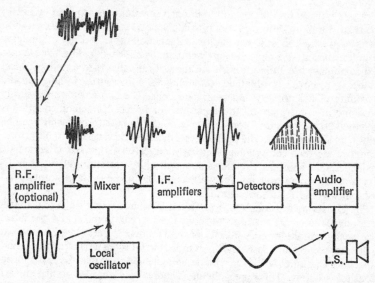

Fig. 116. Block diagram of a superheterodyne receiver showing the waveform at various points in the system.

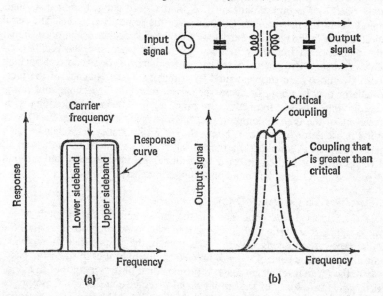

Fig. 117. (a) shows an ideal bandpass response curve for a radio. (b) Since the response is determined mainly by the characteristics of the I.F. transformer we adjust the mutual inductance of the transformer so as to be greater than the critical value. The response is then close to the ideal one.

Fig. 117 illustrates the position. Although transformers are used in T.R.F. receivers, optimum coupling is not possible because the frequency of operation alters as we tune in different stations; the coupling needed for a bandpass response must always be slightly greater than the critical coupling. We see that if the coupling is a little greater than the critical value then an almost ideal bandpass response is obtained. (Coupling is a measure of the magnetic proximity of the primary and secondary coils. The two coils are physically close and are invariably on the same tube or former. Magnetic coupling is increased by having a dust core or ferrite slug within the former which can be moved by a screwdriver. The coupling can therefore be varied.) From the foregoing we see that because superheterodyne receivers use fixed intermediate frequencies the amplifier circuits can operate with maximum stability, sensitivity and selectivity and are not subject to the variable amplification and instability of the T.R.F. receiver.

Fig. 118 shows the circuit diagram of an input and mixer stage of a superheterodyne receiver. The transistor serves the dual role of mixer and oscillator. The biasing conditions are conventional and follow the lines already discussed in the chapter on amplifiers. The input tuned circuit is the ferrite rod aerial and capacitors C_1 and C_t. C_t is the trimmer, and is used to match the two-ganged capacitor sections. The tuned circuit is inductively coupled to the mixer/oscillator. Positive feedback from the collector to the emitter via small inductance windings L_1 and L_2 ensures oscillations the frequency of which is controlled by the tank circuit L_3C_2. C_1 and C_2 are ganged so that a single spindle controls the resonant frequencies of the respective circuits. The aerial circuit tunes in the required carrier wave and L_3C_2 produce oscillations so that the difference in frequency between the carrier wave and the oscillator is always the same irrespective of the carrier frequency. To assist correct tracking both C_1 and C_2 are supplemented by trimmers. The frequency of the local oscillator may be above or below the carrier frequency by the amount of the desired intermediate frequency. In practice, for medium- and long-wave receivers of the type in common use, the intermediate frequency is 465 kHz and the oscillator frequency is above the carrier frequency. The signal is then passed to the I.F. amplifier via the transformer. The output from the second I.F. transformer is then detected by a diode and the resultant audio signal processed by an amplifier in a conventional manner.

Automatic Gain Control (A.G.C.)

A.G.C. is used to compensate for the fluctuations in the strength of the radio signal. It is particularly important in radios used in motorcars since the radio receiver is moving between places where the signal strength may vary considerably. The control signal is taken from the output of the detector stage, where the mean level of the signal is proportional to the strength of the input R.F. signal. Feedback of the output signal from the detector can therefore be used to control the gain of the I.F. amplifiers. Fig. 119 shows the essential circuitry. The signal at the detector output, after I.F. filtering, is fed back as a bias voltage to the base of the transistor acting as the first I.F. amplifier; the network R_1C_1 smooths out the A.F. signal variations. With the diode connec-

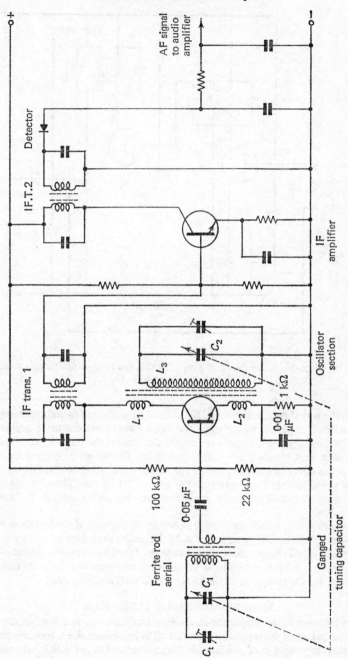

Fig. 118. Circuit diagram of transistor superhet receiver.

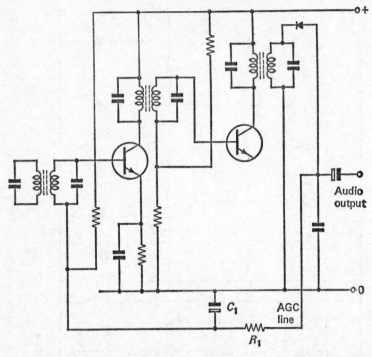

Fig. 119. I.F. and detector stages of a superhet showing how automatic gain control (A.G.C.) is achieved.

ted in the way shown, increases in R.F. signal strength lead to voltage changes at the detector output which cause the mean level to become more negative. The increasingly negative voltage reduces the bias on the first I.F. amplifier's transistor, thus reducing the gain of this stage. The reduction of gain largely offsets the effect of the increases in the aerial signal. Conversely, falls in the aerial signal voltage lead to rises in the gain of the I.F. amplifier. In this way the effect of signal strength variations on the audio output is hardly noticeable.

Unfortunately, a superheterodyne receiver of the type described does not overcome the inherent problems of A.M. reception in respect of atmospheric interference and the electrical interference produced by man-made machines. The way out of this difficulty is to use superheterodyne circuits especially designed for the reception of frequency-modulated carrier waves.

Frequency-Modulated (F.M.) Receivers

On listening to an amplitude-modulated transmission, one is often aware that the quality of the reception leaves much to be desired. A.M. transmissions are inherently capable of yielding good reproduction (e.g. in V.H.F. television receivers), but where bandwidth is restricted, as it must be with medium-wave stations, high-fidelity reproduction cannot be obtained. Also the problem of interference is particularly severe, not only from stations operating on adjacent frequencies, but also from atmospheric electrical phenomena, and man-made electrical disturbances caused by motorcar ignition systems, electric razors, thyristor control equipment, electric welding equipment, etc. Such interference consists largely of amplitude disturbances, and consequently a receiver designed for A.M. reception cannot distinguish between amplitude interference and amplitude-modulated broadcast programmes. Fortunately, amplitude is not the only feature of a carrier wave that can be altered to convey information. We may instead alter the frequency of the carrier wave, keeping the amplitude constant.

If we modulate the frequency of the carrier wave in accordance with the audio signal we create a **frequency-modulated (F.M.) system**. Frequency modulation is now the preferred method of broadcasting high-fidelity sound programmes; in the United Kingdom it is at present the only modulation system that radiates stereo signals. The bandwidth of a broadcast F.M. signal is 150 kHz, which is about ten times that of an A.M. broadcast. We can afford a bandwidth as wide as 150 kHz because the available band space is much greater for F.M. than it is for A.M. broadcasting. This is because the range of F.M. carrier waves is comparatively short (hence interference from other stations is not the problem it is with A.M. receivers), and the F.M. transmitters use carrier waves in the 100 MHz frequency range as opposed to the 1 MHz range. The intermediate frequency is usually 10·7 MHz, so it is not difficult to arrange that the bandpass of an I.F. amplifier is wide enough to accommodate the whole of the audio range up to about 15 kHz.

Apart from the increased fidelity. F.M. reception shows a marked improvement over A.M. systems with respect to noise. With a good aerial system in a location served by a strong station, the background noise of an F.M. receiver is practically undetectable. Such a condition is not possible with A.M. receivers because of atmospheric and man-made electrical interference.

Frequency Modulation and Detection

Fig. 120 shows the frequency-modulation of an R.F. carrier wave by a single audio tone. At the times when the audio tone is passing through its zero value, i.e. at A, C and E, the carrier wave is being radiated at the carrier frequency. At the time (corresponding to B) when the audio signal is at a maximum, the radiated frequency is greater than the carrier frequency, and when the audio signal is at a minimum (corresponding to D), the radiated frequency is less than the carrier frequency.

The **frequency deviation** is the amount by which the frequency varies from the carrier frequency; for example, a carrier frequency may be 100 MHz rising by, say, 40 kHz to 100·04 MHz and falling by the same amount for a sinusoidal audio tone to 99·96 MHz. The deviation in this example is ±40 kHz. The greater the deviation the greater is the strength of the audio signal available after detection. This compares with the A.M. system in which the greater the amplitude change the stronger is the audio signal. Whilst in A.M.

systems the greatest modulation occurs when the carrier amplitude is reduced to zero, we cannot in an F.M. system reduce the frequency to zero. Apart from the impracticability of radiating very low and zero frequency signals, the bandwidth taken by such a transmitter would be so great that no room would

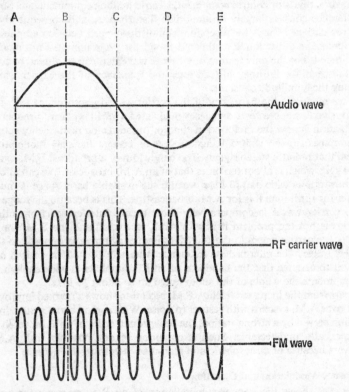

Fig. 120. Frequency modulation of an R.F. carrier wave by a single audio tone.

be left for other transmitters. By agreement the deviation in the United Kingdom, Europe and the United States is limited to ±75 kHz. It will be appreciated that the rate at which the deviations are caused to occur depends upon the audio frequency. For F.M. broadcasting the highest audio frequency used is 15 kHz.

F.M. Detectors

As in A.M. receivers, the function of the detector is to recover the audio information. Because of the nature of frequency modulation, the principle of detection and the associated circuitry differs substantially from the A.M. counterpart. Frequency modulation detection involves two processes.

First, it is necessary to limit any variations in amplitude of the incoming R.F. signal so that a signal which varies in frequency only is obtained. Thus we eliminate all the noise components that have an effect on the amplitude of the signal. This task is carried out by an **F.M. limiter circuit.**

Second, we must convert variations of frequency into an audio output. Although many ingenious methods have been proposed for the detection of frequency-modulated waves, in practice nowadays only one principle is used, i.e. that of the **phase discriminator.** The circuit used is often that proposed by Foster and Seeley and is hence known as the Foster–Seeley phase-shift discriminator. A modification to the phase-shift discriminator known as the **ratio detector** is now in very common use, especially for receivers designed for the reception of both A.M. and F.M. waves. The popularity of the ratio detector is due to the fact that amplitude limiting and detection occur in the same detector, whereas a phase-discriminator needs a separate amplitude-limiter section.

The Phase-Shift Discriminator

The process of converting a frequency variation into a voltage variation is quite complicated, and strictly is outside of the scope of this book. However, readers will naturally wish to know something of the process. The explanation of the detection mechanism requires a knowledge of phasor diagrams for its understanding, but a brief description of vector quantities will be sufficient to clarify the situation.

Force is a vector quantity; this is to say, specifying only the size or magnitude of a force is insufficient. It is necessary to know in addition the direction in which the force is operating. Hence if a force of, say, 10 newtons is operating in a northerly direction, and simultaneously a force of 10 newtons is operating on the same body in a southerly direction, then the combined result of the two forces is zero and the body will not move. We do not say that since two 10 newton forces act on the body this is equivalent to a 20 newtons force. Only if the forces were acting in the same direction would this be so. Similarly, if a force of 10 newtons were acting in a northerly direction and another force of 10 newtons were acting in an easterly direction the resultant force would not be 20 newtons; it would in fact be 14·14 newtons acting in a north-easterly direction. This resultant could be found by drawing. The length of a straight line is proportional to the magnitude of the force and the direction of the line corresponds to the direction of the force. We then draw parallelograms and obtain the combined result of two forces by drawing the diagonal of the parallelogram that passes through the point of application of the forces.

In a similar way we can represent voltages of varying sizes and phases. To do this we use what is called a **phasor.** Phasors have the same mathematical properties as vectors and must be added using the parallelogram rule. They are called phasors (and not vectors) because they represent voltages and currents, both of which are not vector quantities such as force and velocity. Two alternating voltages can therefore be represented by lines the length of which represents the magnitude of the voltage. The angle between the lines represents the phase angle between the two voltages.

Fig. 121 shows the basic circuit arrangement of part of a phase discriminator. The action of the discriminator depends upon the fact that the primary and secondary voltages of a tuned transformer are 90° out of phase at resonance. At any other frequency the phase shift is different from 90° by an amount that depends upon frequency difference between the frequency of the signal and the resonant frequency of the primary and secondary circuits.

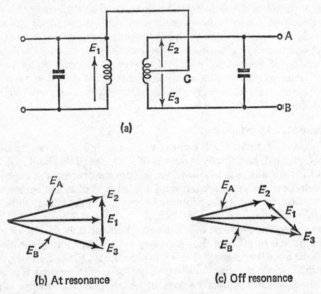

(a)

(b) At resonance   (c) Off resonance

Fig. 121. At resonance, E_1 is 90° out of phase with E_2 and E_3. Since $E_2 = E_3$ in magnitude (but 180° out of phase) E_A and E_B have the same magnitude. Off resonance E_A and E_B are not equal.

Examining the circuit we see that when the carrier wave is not modulated we receive a signal at the carrier frequency. The primary and secondary of the transformer are tuned to this frequency. The voltage between the centre tap of the secondary coil and A is made up of E_2 (the voltage induced in half the secondary) plus E_1, which is the direct connection from the input lead on the primary. At resonance, as previously stated, these voltages are 90° out of phase. We can add E_1 and E_2, but not arithmetically. A line, the length of which represents the magnitude of E_1, can be drawn to the 'east'. Another line representing E_2 must be drawn at right angles to represent the 90° phase shift; this can be drawn to the 'north'. Combining E_1 and E_2 gives E_A as shown. The length of E_A represents the magnitude of the voltage at A with respect to earth. In a similar way we obtain the voltage at B. When the input signal frequency alters, the 90° phase difference between the primary and secondary voltages no longer obtains. The lines representing E_2 and E_3 cannot be drawn at 90° to that representing E_1. One position is shown in Fig. 121(c). We see that E_A is no longer equal in magnitude to E_B. The difference in length depends upon the change of frequency of the input signal. Let us see how we can exploit this in order to obtain an audio signal that depends upon the changes of frequency that occur.

Fig. 122 shows the circuit diagram of the basic phase-shift discriminator. C is merely a blocking capacitor to keep out the d.c. component; it does not

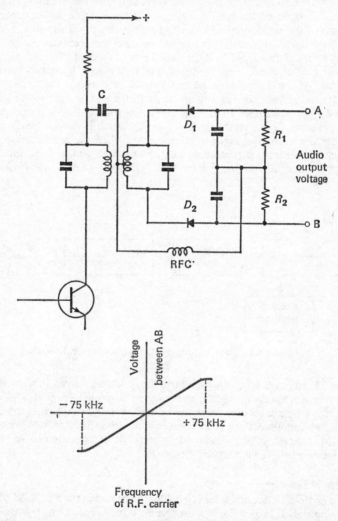

Fig. 122. The phase-shift discriminator. By rectifying and smoothing E_A and E_B, an audio output is produced across AB. When E_A is greater than E_B, A is negative with respect to B. When E_A is less than E_B, A is positive with respect to B.

present any opposition to the signal frequency component. R.F.C. is a radio-frequency choke. It is connected in the line to keep the R.F. component from the audio circuits whilst simultaneously providing a very low impedance path to the audio signals. In the absence of modulation E_A and E_B are equal. The diodes D_1 and D_2 then conduct equally. Voltages are therefore produced across R_1 and R_2 that are equal in magnitude but of opposite sign. The voltage across AB is zero and hence there is no audio output. When the carrier wave is modulated, however, the frequency alters by an amount that depends upon the strength of the audio signal. The stronger the audio signal the greater is the frequency deviation and the greater is the difference between E_A and E_B. When E_A is different from E_B the matched pair of diodes in the discriminator do not conduct equally. The voltages developed across R_1 and R_2 are then unequal and the difference is the audio signal produced at the terminals A and B. The polarity of the voltage at A varies with respect to that at B depending upon whether D_1 or D_2 is conducting harder. This in turn depends upon whether E_A is greater or less than E_B, which in turn depends upon whether the actual frequency is greater or less than the centre carrier frequency. We see from the characteristic that a properly designed discriminator produces an audio output voltage that is proportional to the frequency deviation.

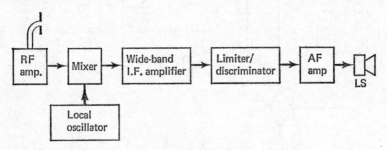

Fig. 123. Block diagram of an F.M. superheterodyne receiver.

Fig. 123 shows in block diagram form the system used in an F.M. receiver. Such receivers always use the superheterodyne principle because it is the only satisfactory type of receiver that can amplify the weak aerial signals to a level where their amplitudes can be maintained constant by the limiter. We see that an F.M. receiver is very similar in its principle of operation to an A.M. receiver except in the limiter-detector stage.

The Ratio Detector

The phase-shift discriminator is sensitive to changes in the amplitude of the incoming signal. In practice it must therefore be preceded by an amplitude limiting stage. In an attempt to eliminate the necessity for a limiter, a modification to the phase-shift discriminator, known as the **ratio detector**, has been devised. The basic circuit, shown in Fig. 124, is similar to the phase-shift discriminator circuit of Fig. 122, but it will be seen that diode D1 has been reversed and that the audio signal is taken from one of the capacitors. Because of the reversal of the diode D1, both diodes conduct on the same R.F. half-cycle. In the absence of modulation each diode conducts equally and the voltages developed across capacitors C1 and C2 are equal. When the carrier is modulated, however, these voltages are not equal for the same reasons as were

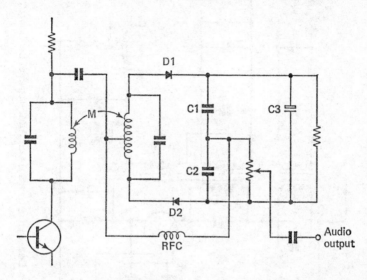

Fig. 124. Circuit diagram of ratio detector. M represents the mutual coupling between the primary and secondary coils.

given in the explanation of the action of the phase-shift discriminator. Because of the presence of the large capacitor C3, however, the sum of the voltages across C1 and C2 is kept constant regardless of the ratio between them. Since this sum remains constant, amplitude variations cannot occur. The ratio between the voltages across C1 and C2 changes continuously in accordance with the instantaneous frequency of the carrier wave.

A modern version of the circuit is shown in Fig. 125. This arrangement is very popular in sets intended for use on both A.M. and F.M. wavebands. (The A.M. section is shown in dotted lines.) The principle of operation is similar to that for the circuit of Fig. 124. No direct connection is made to the I.F. amplifying transistor, however. The voltage (equivalent to E_1 for the phase discriminator) is developed across another secondary coil, L_3, and is added to the centre-tap voltage of coil connected to the diodes. As with the circuits of Fig. 122 and 124, deviations of the carrier frequency from the centre frequency result in unequal conduction of the detector diodes. Difference voltages result in an amplitude-varying audio-frequency signal across L_3. R_1 and C_1 form an R.F. filter.

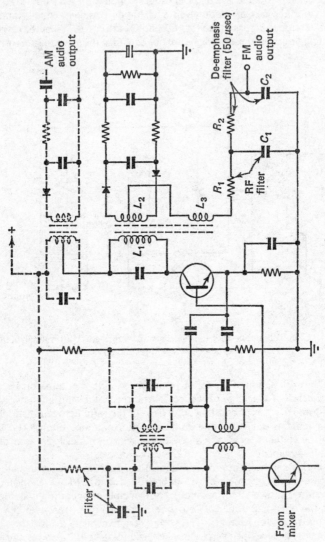

Fig. 125. A.M./F.M. radio receiver Dotted components are the A.M. section of the receiver.

R_2 and C_2 make up a 'de-emphasis' filter. Such a filter is common to all F.M. receivers. (It has not been shown in the circuits of Figs. 122 and 124.) This filter is used to reduce the audio signals at high frequencies. The amount of reduction increases as the frequency increases—in other words, the higher audio frequencies are 'de-emphasized'. The de-emphasis filter thus compensates for the deliberate boosting of the high frequencies at the transmitter. Such a boost is given in order to improve the signal-to-noise ratio. When high-frequency boost is applied, great care is taken to minimize the noise content. At the receiver end we de-emphasize the signal to obtain the correct acoustic balance and in doing so we also reduce any noise introduced by the receiver circuitry. The signal-to-noise ratio is thus improved. The time constant of the filter must match that of the transmitter; by general agreement this time constant is 50 microseconds in European countries.

Summary

The basic sections of a radio communication system are: (a) a radio **transmitter** to generate modulated radio-frequency waves and (b) a radio **receiver** to select and amplify the desired signal and to demodulate it so as to recover the audio information.

Modulation is the process of superimposing information on a radio-frequency carrier wave. These waves may be amplitude modulated or frequency modulated. **Amplitude modulation** gives rise to sidebands; the bandwidth of the channel must therefore be twice the band of frequencies included in the modulation. The deeper the modulation the stronger is the audio signal; overmodulation results in distortion. In **frequency modulation** the greater the audio signal the greater is the deviation of the frequency from that of the carrier wave; the audio frequency determines the rate at which the deviations of frequency occur. The greatest deviation used in practice is ± 75 kHz and the maximum audio frequency broadcast is 15 kHz.

Electromagnetic (radio) waves may travel from transmitter to receiver as **ground** or **surface waves**, as **ground-reflected waves**, **direct** or **line-of-sight waves** or **sky waves**.

A radio receiver must have: (a) an **aerial** to convert the radio waves into electrical signals; (b) **tuned** LC **circuits** to select the desired signal; (c) a **detector system** to recover the audio information; (d) an **audio amplifier**; and (e) a means of reproducing the original sound. A **T.R.F. receiver** consists of one or more stages of R.F. amplification followed by a detector and audio amplifier. The **superhet** converts all received carrier frequencies to a fixed intermediate frequency and thus attains maximum sensitivity, stability and selectivity. The I.F. amplifiers are followed by a diode detector, audio amplifier and loudspeaker.

Frequency-modulated receivers use the superheterodyne principle. For the demodulation of F.M. waves a **phase discriminator** or **ratio detector** is required. Apart from this the F.M. circuitry is similar to that used for A.M. sets.

Questions

1. Why is modulation necessary for the transmission and reception of sound programmes?

2. What factors influence the choice of A.M. or F.M. in the reception of domestic entertainment programmes?

3. Sketch the waveform of an R.F. carrier wave amplitude-modulated to a depth of 70 per cent.

4. What is the function of a limiter in a frequency-modulated receiver?

5. Why do F.M. transmitters use much higher carrier frequencies than transmitters of A.M. waves?

6. Why is it necessary to use crystal-controlled oscillators in radio transmitters?

Suggested Further Reading

King, G. J., *Beginners Guide to Radio*, Newnes–Butterworth, 1973.
Ness, H., *Handbook of Tested Transistor Circuits*, Babani Press, 1971.

10

THE CATHODE-RAY OSCILLOSCOPE

The modern cathode-ray oscilloscope is designed as a measuring instrument, and is the most useful of all electronic test instruments. So great is its versatility that workers in every branch of scientific activity now find the instrument to be almost indispensable. If limited to the purchase of a single item of electronic measuring equipment, the majority of experienced workers would select an oscilloscope.

The instrument delivers its information in the form of a graph or trace on the screen of a cathode-ray tube. By giving an immediate visual display of the amplitude and waveform of the quantity under consideration, a rapid insight is gained into the functioning of electronic and electrical circuits. Where the phenomena to be studied are not electrical—for example, the mechanical vibrations in beams and machinery, the variations of pH in a solution, temperature fluctuations, patients' heart-beats, etc.—then all that is necessary is to find suitable transducers that give output voltages that correspond to the variations of the quantities involved. Strain gauges, piezo-electric crystals and moving-coil microphones, photoelectric cells and thermocouples are examples of commonly used transducers. The electrical output of a transducer is applied to the input terminals of the oscilloscope, and the latter automatically draws out the waveform on the screen. The displaying of waveforms constitutes an important advantage over the usual moving-pointer instruments, since the latter can yield only amplitude information.

The inertia of the moving parts of such electromechanical systems as moving-coil meters, potentiometric and other pen recorders, etc., imposes a severe limitation on the maximum signal frequency to which such instruments can respond. The oscilloscope, on the other hand, is a truly electronic instrument having no moving parts except the beam of electrons, which is, for all practical purposes, inertialess. The cathode-ray tube can thus respond to very rapid alternations of voltage. General-purpose laboratory oscilloscopes having adequate responses up to 30 MHz are now common, and special-purpose instruments are available that can respond to signal frequencies as high as 1,000 MHz.

The oscilloscope may be considered as a combination of a cathode-ray tube together with appropriate amplifiers and power supplies. In addition a special relaxation oscillator, known as a timebase, is incorporated.

The Cathode-Ray Tube

This tube is the major device in an oscilloscope. It consists of a source of electrons, focusing electrodes, a deflection system and a fluorescent screen, all contained in a suitably shaped glass tube. It is illustrated diagrammatically

203

in Fig. 126. The tube is highly evacuated to pressures of the order of 10^{-6} torr (mm Hg). The aim within the tube is to produce a narrow beam of electrons that comes to a sharp focus at the screen.

The **screen** is a thin layer of phosphor that glows at the point of bombardment of the electrons. The phosphor, binding chemicals, and activators to increase the luminous efficiency, are deposited on the inside of the face of

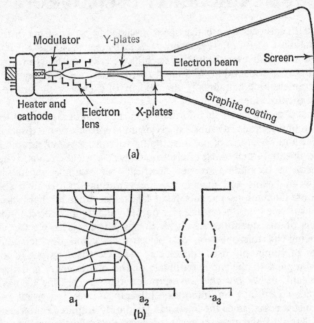

(a)

(b)

Fig. 126. (a) Diagrammatic representation of a cathode-ray tube showing the main electrode features. (b) The electric field within the three-anode electron lens system. The solid lines show the shape of the electric field, and the broken lines show the position of the equipotential surfaces. The similarity between the shape of the equipotential lines and that of converging glass lenses should be noted.

the tube. The type of phosphor used determines the colour and persistence of the trace; for viewing work yellowish-green traces are used. This colour is obtained by using zinc orthosilicate with small amounts of manganese as an activator. **Persistence** is expressed as short, medium or long, and refers to the length of time taken for the trace to become invisible after the electron beam has been removed. Medium persistence (i.e. where the brightness of the trace is reduced to about 1/1000th of its original value in approximately 50 milliseconds) is adopted for viewing tubes. Blue, short persistence traces are used for photographic work since the sensitivity of film emulsions used in oscillograph recording is usually highest in the blue region of the spectrum. Orange and yellow long-persistence traces are used in radar equipment and in some large demonstration oscilloscopes.

The electrons are produced by heating a small area of barium and strontium oxides deposited on a nickel surface. The beam current is controlled by the **modulator**, i.e. a nickel tube that is coaxial with the cathode. The modulator is maintained at some potential which is negative (about -15 to -100 V) with respect to the cathode. As the potential is made increasingly negative, fewer electrons reach the screen and thus the trace becomes fainter. Eventually, by applying a sufficiently high negative potential, the beam current is reduced to zero and the tube is then said to be 'cut off'. The knob on the front of the instrument that is used to control the modulator voltage is marked 'brightness' or 'brilliance', since by operating this control the brightness of the trace on the screen may be varied. Because of the similarity between the action of the modulator in a cathode-ray tube and the grid in a thermionic valve, the modulator is frequently called the **grid**, in spite of its physical shape.

The electrons are accelerated towards the screen and simultaneously brought to a sharp focus by a system of three electrodes called **anodes** because they are held at positive potentials. The three anodes form the electron lens. These anodes, a_1, a_2 and a_3, consist of a system of cylindrical electrodes, arranged in line coaxially with the modulator and neck of the tube. The anodes a_1 and a_3 are connected together electrically within the tube and are maintained at a high positive potential relative to the cathode. Different tubes require different anode potentials, but values commonly used lie between 1 kV and 5 kV. Bewteen a_1 and a_3 there is a focusing electrode a_2. The voltage on this electrode may be varied by operating the focus control on the front panel of the instrument. This voltage is commonly one sixth to one third of the final anode voltage. The difference in potential between a_2 and the other anodes, a_1 and a_3, creates an electric field through which the electrons must travel. The electric field configuration is such that the electrons are made to converge to a point. The action of the electric fields on the electrons is analogous to the action of glass converging lenses on beams of light, which is why the anode system is referred to as an **electron lens**. The adjustment of the potential on a_2 by the focus control alters the focal length of the lens system. With the correct adjustment the electron beam is brought to a sharp focus at the screen. The heater, cathode, modulator and anode system are collectively referred to as the **electron gun**.

Deflection System

In order to make the electron 'pencil' draw on the screen, some suitable method of deflecting the electron beam must be employed. In nearly all cases of oscillographic work the deflections are produced by electric fields. The electric fields are created by applying voltages, proportional to the quantities to be represented, to two pairs of **deflector plates**, arranged mutually at right-angles and between which the beam passes. When one plate of a given pair is made positive with respect to the other, the negative electrons are accelerated towards it. The beam may therefore be deflected in a direction at right-angles to the given pair of plates. Since two pairs of plates are provided, the beam may be made to scan the whole of the useful screen area. The pair nearer the anodes has a greater effect on the beam; therefore this pair is reserved for the signal. Following the usual convention in graphical work, the deflections produced are in the Y-direction. These deflectors are therefore called **Y-plates**. The other pair are called **X-plates** because they produce deflections in the X-direction. The plates diverge at the ends nearest the screen to prevent the beam from being interrupted when large deflections are required.

The deflection produced on the screen is directly proportional to the deflecting voltage, a fortuitous arrangement that yields a linear scale for the axes. In practice, the spot diameter will limit the smallest deflection that can be detected. The resolution of the cathode-ray tube may be defined as the voltage applied to the deflector plates that produces a deflection equal to one spot diameter.

Double-Beam Tubes

It is frequently an advantage to have two quantities displayed simultaneously on the screen as separate graphs. This may be achieved by creating two beams of electrons within the tube. The single beam may be split by a beam-splitting plate or alternatively two electron guns can be used. Each beam passes through its own Y-plates, but both beams pass through a single pair of X-plates. The deflections in the X-direction are therefore common to both traces.

Electronic switching can be employed to give two traces with a single-beam tube, this method being favoured by some manufacturers. An electronic beam-switching circuit consists of a pair of identical amplifiers that are controlled by a square-wave generator. The outputs of the amplifiers are connected together and fed to the input of the single-beam oscilloscope. Each amplifier input receives one of the signals to be examined. The square-wave generator alternately biases the amplifiers so that whilst one amplifier is operating the other is cut off. When the square-wave switching frequency is much greater than either of the signal frequencies, portions of each signal are alternately presented to the input socket of the oscilloscope. The resulting trace is then composed of portions of each waveform, but because of the high switching frequency an illusion of two separate traces is obtained. Variable bias voltages are applied to the amplifier and square-wave generator so that the two signal traces can be separated for easier examination.

Post-deflection Acceleration

Tubes that have been designed to achieve the brightest possible trace on the screen without affecting unduly the deflection sensitivity are known as post-deflection acceleration (p.d.a.) tubes. The deflection sensitivity of an ordinary cathode-ray tube is inversely proportional to the final anode voltage, and thus brightening the trace by the use of greater accelerating voltages reduces the sensitivity. Higher deflection voltages are then required when producing a given deflection with a brighter trace, and this entails more costly amplifiers to operate the deflector plates. However, it has been discovered that if the main acceleration of the electrons in the beam is effected after deflection, the reduction in sensitivity is not very great; p.d.a. tubes therefore have an internal ring or graphite spiral on the glass wall near to the screen; this final accelerating electrode is maintained at a very high voltage (about 10 kV in many cases) relative to the cathode.

When electrons strike the screen the latter would normally acquire an undesirable charge. This is prevented by collecting the secondary electrons emitted by the screen after bombardment by the primary beam electrons. The collection is effected in p.d.a. tubes by the final graphite electrode. In other tubes, a simple graphite coating on the walls near the screen is connected to a_3, and thus has sufficient potential to collect the secondary electrons.

The Oscilloscope

A complete oscilloscope is shown in block diagram form in Fig. 127. It is necessary to have power supplies in order to operate the cathode-ray tube. These supplies provide the heater currents and high voltages necessary to operate the tube electrode system and the remaining electronic circuitry. Since signal voltages are rarely large enough to produce satisfactory deflections when applied directly to the deflector plates, intermediate X- and Y-amplifiers must be used. A timebase produces the time axis along the horizontal or X-direction in the way described below.

The Timebase

The majority of phenomena studied with an oscilloscope are periodic—that is to say, the variations are repeated exactly in equal successive intervals of time. All oscilloscopes therefore include a special type of relaxation oscillator and this, together with the X-amplifier, constitutes what is known as a **timebase**.

The function of the timebase is to deflect the beam in the X-direction so that the spot moves from left to right across the screen with uniform speed. Having travelled from left to right, the spot must then be returned to its original position as rapidly as possible (ideally in zero time). The output waveform of the timebase should therefore approximate very closely that shown in Fig. 128. From the diagram it can be deduced that the distance travelled horizontally by the spot must be directly proportional to time so that, in effect, the horizontal axis is a time axis. If simultaneously the signal voltage is applied to the Y-plates, the spot delineates the signal waveform.

To be effective for viewing purposes the timebase frequency must be that of the signal or a submultiple of it. When this condition obtains, successive traces coincide and, because of the viewer's persistence of vision and the tube persistence, an illusion is created of a stationary pattern. The timebase frequency is varied by adjusting two controls at the front of the instrument. One control selects the approximate frequency and is often called the time control, and the other, the fine frequency control, adjusts the oscillator frequency to the desired value. The stability of modern timebase generators is good enough to enable manufacturers to abandon the former practice of marking the controls in terms of the sweep repetition rate. Instead these controls are marked in time units specifying the sweep rate. The coarse frequency control that selects the size of the capacitor in the main charging

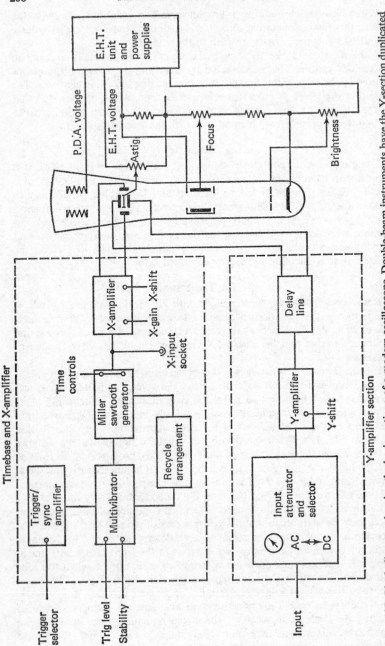

Fig. 127. Block diagram showing the basic sections of a modern oscilloscope. Double-beam instruments have the Y-section duplicated and use a beam-switching circuit or a double-beam tube.

circuit of the timebase oscillator is now marked in time units, and the fine frequency control acts as an interpolator between these fixed ranges. The sweep rate can then be read off directly. Thus, if from the controls it is known that the sweep rate is 50 ms/cm, then a phenomenon producing a trace that extends over say 3 cm in a horizontal direction has a duration of 150 ms. The calibration of the timebase in time units is particularly helpful when examining pulse waveforms.

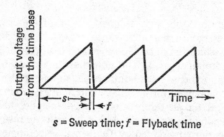

s = Sweep time; f = Flyback time

Fig. 128. The waveform required from an ideal timebase. Strictly, f should be zero, but in practice we must accept a small finite value where $f \ll s$.

The locking of the timebase frequency to that of the signal is termed **synchronization**. This is essential since the frequency of the timebase, and often that of the signal, is not quite constant; consequently, if the timebase were free running, successive traces would not coincide and a confused screen pattern would result.

When considering the purchase of an oscilloscope, it is important to know what type of synchronization is used. In the older 'recurrent' types still found on some of the cheaper models, synchronization is achieved by feeding a small fraction of the signal voltage directly into the timebase. (This fraction may well come from the Y-plate voltages.) This has the effect of firing the timebase oscillator at the right time to ensure coincidence of successive traces. The frequency of the timebase must be carefully adjusted to be marginally slower than the signal frequency, and positive locking is not always easy to achieve.

A much better system found in all good-quality oscilloscopes uses a triggered system of synchronization. In this way the input signal exerts a more positive control over the repetition frequency of the timebase. The trigger selector switch often has six positions. The $+$ and $-$ internal positions mean that the trigger impulse is obtained from the Y-amplifier, the sign determining whether the sweep starts as the signal waveform is going positive or negative. The external positions have a similar significance except that the trigger signal is obtained from external sources. The alternative positions ensure synchronization with the mains frequency, the trigger voltage usually coming from one of the heater supplies. This is often useful when a determination of signal frequency is to be made, this frequency being not much higher than ten times the mains frequency; triggering from the mains is also useful when studying phenomena associated with machinery that is mains operated.

The timebase oscillator depends upon the controlled charging and discharging of a capacitor of the appropriate size. Different sizes are selected by a control on the front of the instrument. It will be recalled that the voltage across a capacitor is proportional to the charge stored. By charging the capacitor via a constant-current circuit, the charge, and hence the voltage, will increase proportionally with time. Special circuitry is employed to start and stop the charging at the appropriate time.

Y-amplifiers

There can be no doubt that a consideration of the specifications of the Y-amplifier is of paramount importance before the purchase of any oscilloscope. The classification of oscilloscopes and their cost depend largely upon the type of Y-amplifier used. The potential user must therefore consider carefully his requirements since failure to do so may lead to the purchase of an instrument inadequate for his needs. Alternatively, money may be wasted in buying facilities that are rarely or never used. If a potential user is uncertain of his future requirements a good solution is to buy an instrument with plug-in units, such as those marketed by Tektronix, Telequipment and Solartron. A basic oscilloscope sufficient for his immediate needs may be bought at reasonable cost, and, if later work demands a higher specification, it will be possible to buy the necessary plug-in units to extend the facilities of his instrument.

The cost of the Y-amplifier section, and hence the oscilloscope, depends largely on the facilities required. Factors to consider are bandwidth, sensitivity, rise time, transient response and delay facilities.

The Y-amplifiers themselves are conventional enough and depend nowadays on the excellent properties of available integrated circuits. It is still necessary, however, to use conventional transistors (or even thermionic valves) to produce the high final voltages to operate the Y-plates. Facilities are also provided for varying the bias voltages in the amplifier so that shifts of the trace in the Y-direction are obtained.

Practical Considerations

In spite of the seemingly complex array of knobs and dials on the front of many laboratory oscilloscopes, no alarm need be experienced in using the instrument. Apart from one exception, no damage can result from incorrect settings of the controls. The handbook for the instrument should be available and consulted before putting the machine into operation. This will explain the function of each control and how to make adjustments to obtain satisfactory operation. The exception mentioned above relates to the **brightness** or **brilliance control**. In operation, this control must be adjusted to give a trace with the minimum brightness consistent with satisfactory viewing; the risk of burning the screen is then eliminated. Stationary spots within the viewing area should also be avoided to reduce the risk of burning. The timebase should be kept running and the spot drawn out into a line. If this is not possible then a **shift control** should be used to move the spot to the edge of the screen where screen defects are of relatively minor importance. An increase in **contrast** can be obtained by operating the oscilloscope in a position which avoids direct light falling on to the screen. Visors or hoods are useful in this respect; coloured filters in front of the screen are also used to increase the useful contrast. For quantitative observations, it is useful to have the filter ruled with a graticule. A convenient and much used arrangement involves the ruling of the main axes, suitably subdivided, together with horizontal lines spaced at intervals of one centimetre.

When an oscillogram is being displayed, it is necessary to adjust the brightness, focus and astigmatism controls until the best picture is obtained. 'Astigmatism' is an optics term which means an inability to focus correctly simultaneously in two directions. In an oscilloscope these directions are the X- and Y-axes. The defect is minimized by ensuring that the mean voltage on the Y-plates is correctly adjusted; this is the function of the astigmatism control. For the best focus, it is necessary to adjust the focus and astigmatism controls together since they are not independent in their action on the trace. For periodic waveforms the timebase should be set so as to give three or four complete cycles of the waveform. The last waveform will not be quite complete owing to the finite time it takes for the spot to return to the origin. On most modern oscilloscopes the return is not shown on the screen because a negative blanking pulse, applied to the modulator during flyback, cuts off the beam.

Failure to obtain a trace on the screen is common with newcomers to the instrument. Apart from ensuring that mains power is being delivered to the instrument (usually by observing the pilot light), users should check that the brightness control is sufficiently advanced to make a trace possible. If the X-shift or Y-shift controls are not correctly adjusted, sufficient bias voltage may be being applied to the X- or Y-plates to deflect the spot off the screen altogether. With the X- and Y-shift controls in approximately their mid-positions the continued absence of a trace is then probably due to incorrect settings of the stability or level controls.

It is advisable to check the calibration of the Y-deflections and the timebase at regular intervals throughout the life of the instrument. Minor adjustments are possible from outside the case, but, unless the operator is skilled at repairing electronic instruments, it is best not to delve into the oscilloscope.

Voltage Measurements

Having calibrated and checked the attenuator settings associated with the Y-deflections, the application of the signal produces deflections from which the input signal voltage can be easily determined. Ideally calibrations should be made immediately prior to making a measurement; errors arising from changes in amplification with time due to fluctuations in supply voltages, ageing of components, etc., are thus eliminated. Due regard must be paid to two further errors. The first error arises because of non-linearities of the Y-amplifier and cathode-ray tube. If these are appreciable then strictly a calibration voltage should be of a similar amplitude to that of the signal voltage to be measured. The second error relates to the frequency response of the Y-amplifier. Since this response is never perfect the frequency of the calibration voltage should be similar to the frequency of the voltage to be measured.

Some oscilloscopes have their shift controls calibrated in volts. Thus by shifting the trace relative to a fixed horizontal line it is easy to read off the voltage difference between any two points on the waveform. The accuracies attainable with this method are not usually very high.

The calibration of the Y-amplifier and the associated deflection on the screen can most conveniently be made by using the calibration voltage available from the instrument itself. The clipping of a sine wave with zener diodes produces a square wave of accurately known amplitude; many manufacturers are now using this method to provide a calibration unit.

The timebase can easily be checked by building or buying a crystal-controlled oscillator. A 1 MHz oscillator provides one cycle in every microsecond. By using a chain of dividers as in Fig. 129, we obtain the submultiple

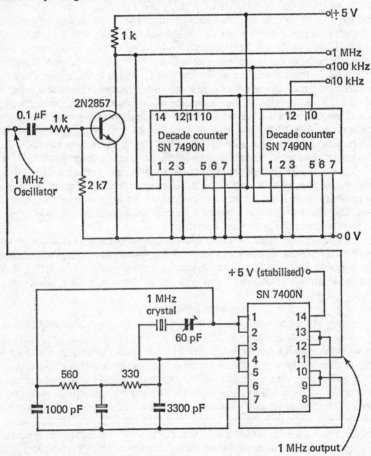

Fig. 129. Method of obtaining accurately known frequencies from a crystal oscillator drive.

frequencies of 1 MHz. The multivibrators are locked to the crystal oscillator, and hence the calibration frequencies are very accurately known.

Frequency Measurements

Frequency measurements can be made very elegantly and accurately by means of a **Lissajous figure** if sinusoidal voltages are involved. The signal voltage of the frequency it is desired to measure is applied to one set of deflector plates, via the appropriate amplifier, and a sinusoidal voltage of accurately known frequency is applied to the other set of deflecting plates, again via the appropriate amplifier if necessary. The timebase must be switched off by turning the stability control fully anticlockwise. The resulting traces on the screen are known as Lissajous' patterns, so called after Professor Lissajous, who first produced them by combining at right-angles the component oscillations of two tuning forks; the combination in his case was effected by optical means. By studying the shape of the pattern an accurate comparison of the frequencies can be made. For any given frequency ratio many patterns are possible, depending upon the relative phases of the two voltages.

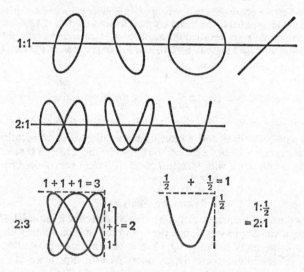

Fig. 130. Typical Lissajous' patterns (for the significance of the figures and the estimation of the frequency ratios see text).

Fig. 130 gives some possible Lissajous' patterns. If two mutually perpendicular and centrally placed axes are drawn through the Lissajous figure, the latter crosses the y-axis n times and the x-axis m times. Where a curve crosses itself on an axis, two intersections on that line are counted. The frequencies of the sources producing the horizontal and vertical deflections are in the same ratio as the numbers of times the pattern is intersected by the y- and x-axes, i.e. $f_y/f_x = m/n$.

Phase Difference Measurements

The phase difference between two sinusoidal waveforms can be measured using the arrangement of Fig. 131. We are in effect using a Lissajous figure that indicates a 1 : 1 ratio. In applying the formula we must ensure that the

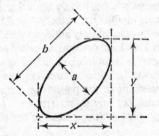

Fig. 131. When the phase shift in a network or amplifier is required, the input and output voltages can be fed to the X and Y inputs to produce the 1:1 Lissajous figure. The gain settings must be such that the amplitudes of the voltages presented to the deflector plates produce a pattern in which $x = y$. The phase angle φ is then given by

$$\text{Tan } \frac{\varphi}{2} = \frac{a}{b}$$

where a and b are the minor and major axes of the ellipse and φ is the phase angle.

amplitudes in both the x and y directions are adjusted to be identical. The phase shift in each amplifier channel must also be the same, so that wherever possible an oscilloscope with identical X- and Y-amplifiers should be used.

Waveform Analysis

Perhaps the most frequent use made of the oscilloscope in most laboratories is the general viewing of waveforms present in the various items of electronic equipment being used. By knowing what waveforms are obtained when the apparatus is functioning correctly, an operator with experience can quickly diagnose many faults when viewing the patterns obtained from apparatus that is not functioning correctly. It is advisable to build up from memory and records a set of oscillograms associated with newly bought equipment known to be operating successfully. Such records then supplement the manufacturer's data regarding the fixed voltages to be expected at various points in the circuits of the apparatus. Many of the circuits in this book yield waveforms that should be examined by an oscilloscope and recorded for future reference. Figs. 132 and 133 indicate by way of example the waveforms to be expected with various circuits.

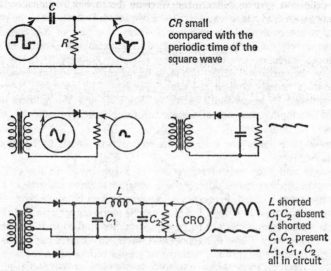

Fig. 132. Typical traces to be expected when examining the above circuits with a cathode-ray oscilloscope.

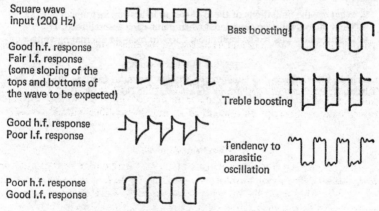

Fig. 133. Square-wave testing of an audio amplifier.

Summary

The cathode-ray oscilloscope is designed as a measuring instrument that displays signal information in the form of a graph or trace on the face of a cathode-ray tube.

The **cathode-ray tube** consists of an electron gun which produces and focuses a thin beam of electrons onto a fluorescent screen. The latter glows at the point of impact of the electrons.

A **deflection system** deflects the electron beam electrostatically in accordance with the voltage waveform to be displayed. Deflections in the vertical direction are controlled by applying voltages to a pair of **Y-plates**. These voltage waveforms are those of the signal to be examined. Deflections in the horizontal direction are controlled by applying voltages to the **X-plates**. Usually, the horizontal deflections must be proportional to time because the phenomena usually studied are time-dependent. The necessary X-voltages are produced by a saw-tooth oscillator. Both X- and Y-amplifiers are required to amplify the control signals before application to the X- and Y-plates.

Double-beam tubes are available to enable two signals to be displayed simultaneously.

Voltage and **time measurements** can be made by using the calibrated controls of the instrument. Alternatively, separate calibrated signals of known magnitude can be used. **Frequency measurements** can be made by observing Lissajous' patterns. For these the time base is made inoperative and the two signals whose frequencies are to be compared are applied to the X- and Y-inputs respectively. **Phase differences** and **waveform analyses** can be made by a proper interpretation of the patterns or graphs produced on the screen.

Questions

1. Why is the cathode-ray oscilloscope the most useful of all measuring instruments?
2. What are the limitations of the CRO as an indicating instrument?
3. What factors must be considered when purchasing an oscilloscope?
4. Describe how you would calibrate a CRO to measure alternating voltages.

Suggested Further Reading

Czech, J., *The Cathode Ray Oscilloscope*, Philips Technical Library, 1957.
Turner, R. P., *Practical Oscilloscope Handbook*, Iliffe Books Ltd., 1964.

11

TELEVISION

'If u cn rd ths u cn gt a gd jb as a shthnd typst.'

Most readers will be acquainted with messages of this type. The example given above is very similar to that which appears on advertising posters in London's underground trains. Such a message is capable of interpretation even though a good deal of the written information is absent. Once people have acquired certain basic reading skills, they can supply the missing information and hence interpret the message.

This ability is not confined to reading the printed word. Both the ear and the eye can 'supply' missing information that comes in the form of aural or visual messages. In telephone communications, for example, we know that conversations are transmitted with a restricted frequency range and with a good deal of distortion. In spite of this, however, there is usually no difficulty in interpreting the message, although since the ear is a sensitive organ, the messages sent to the brain are never interpreted as being the authentic sound. As we shall see in the next chapter, deceiving the ear/brain combination into thinking that it is listening to the true sound is extremely difficult.

The eye is somewhat easier to deceive. It receives its information in the form of light pictures that are focused by a lens onto the inner back surface of the eye, called the retina. This retina is made up of thousands of rods and cones which terminate in light-sensitive nerves. After exposing the retina to light it takes about one tenth of a second for the eye to readjust in readiness for any change in the scene. This persistence of vision enables us to deceive the eye into sensing motion where none is present. If still pictures are presented to the eye at a rate greater than ten per second the eye cannot perceive them separately; the pictures then merge into a continuous impression. Small changes in the position of objects in successive pictures are then interpreted as moving objects by the viewer.

As readers will know, the cinema depends upon this principle. Still pictures are screened at a fast rate and the illusion of a moving picture is created. In cinema jargon each still picture projected is called a 'frame'. It has been found that although an impression of motion is created when the frame projection rate is about 15 per second, a flicker effect, which is most irritating, is experienced. (Hence is derived the slang word 'flicks' for cinemas in the early days.) In cinemas today 24 frames are flashed onto the screen per second. Even at this rate some flicker would be evident were it not for the fact that modern projectors are fitted with a shutter that breaks the projection of each frame into two equal periods. We thus see each still picture twice, but the effective projection rate has now been increased to 48 per second. At this rate all traces of flicker are eliminated.

Whilst the persistence of vision makes possible the creation of an illusion in the cinema, this factor alone is insufficient to allow the transmission and reception of television pictures to take place. We need in addition to exploit another property of the eye.

Normally, when we view a scene the picture that is focused on the retina consists of a continuous variation of light and shade, colour and hue. It is not possible to convey an infinite variety of light and shade impressions to the brain, however, because of the limited number of light-sensitive nerve endings in the retina. Consequently, if a change in light intensity takes place over the area of a single nerve ending, the eye cannot detect the change since a single nerve ending can transmit only a single response to the brain at any one time. This property of the eye is called 'acuity', which is concerned with the sharpness of vision. If readers have ever looked closely at a newspaper photograph, or examined it under a magnifying glass, they will realize that the picture is made up of thousands of tiny dots of varying size. At normal viewing distance, however, the eye cannot resolve these as separate dots and the illusion is created of a picture with a continuous change of light and shade. This property, together with the persistence of vision of the eye, makes it technically possible to transmit and receive television pictures.

Fig. 134 shows a system of horizontal lines, each of which has been thickened in a special way. If such an arrangement is viewed from a distance, the

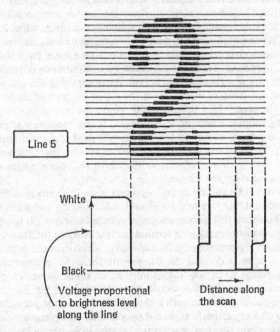

Fig. 134. Line structure of pattern representing a shaded figure 2 together with the voltage representing the instantaneous 'whiteness' level along line 5.

eye is not able to resolve the separate lines very well and there is no difficulty in recognising the picture of the figure '2' followed by a dot. The illusion of a 'solid' figure '2' would be improved if the number of lines in the area were increased.

We see now how such information could be transmitted. The television transmitting station must first break down a given picture into a line structure and develop an electrical signal that corresponds to the instantaneous brightness level at any point on a line. Fig. 134 shows how the waveform could appear for one of the lines in the picture. The brightness level can then be made to modulate the amplitude of an R.F. carrier wave, which subsequently can be transmitted as a television signal. The television receiver must, on receiving the signal, be able to reassemble the line information in such a way that a coherent picture is formed. Furthermore, in order to create the illusion of movement when the scene to be transmitted depicts motion, it is necessary to transmit complete picture frames at a rate that enables us to make use of the persistence of vision of the eye. Each frame must be broken down into a large number of lines if the final picture is to contain an acceptable amount of detail. If too few lines are used the viewer would be irritated by a 'liney' picture and the illusion of a well-defined moving picture would be spoilt. In the United Kingdom pictures based on a 405-line system were initially used, but as the electronic techniques were developed and improved the number of lines was increased to 625. In the United States a 525-line system is used.

A complete television system is shown in block diagram form in Fig. 135. We see that in principle at least the transmitting and receiving circuits follow the lines already described in the chapter on A.M. and F.M. radio systems. For television purposes additional circuitry is required to supply the necessary scanning, synchronization and deflection arrangements.

The Camera Tube

The first requirement at the transmitter is some means of converting the visual picture into a corresponding electrical signal. This is achieved in the camera tube. Several tubes have been invented for the purpose, but the **image orthicon** is the type mainly used for black and white work. The popularity of this tube is due to its very high sensitivity.

Fig. 136 shows in principle the construction of the image orthicon. The scene to be viewed is focused by a camera lens onto the flat end of the tube. On the inner surface of this flat end is a semi-transparent photocathode which emits electrons from those portions which are illuminated. The number of electrons emitted from any small area depends upon the intensity of the illumination falling on that area. These electrons are accelerated towards a target plate by means of a tube known as an **accelerator grid**, which is maintained at a positive potential. On striking the target these electrons knock further electrons from the material of the plate at the point of impact. These secondary electrons are then collected by a fine-mesh wire screen placed close to the target on the photocathode side. The target plate is made from special material and is very thin. The region from where the secondary electrons are

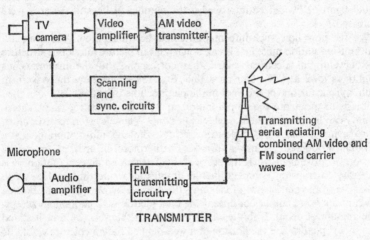

TRANSMITTER

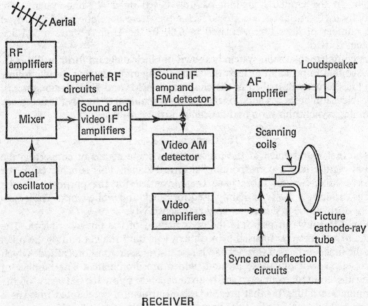

RECEIVER

Fig. 135. Simplified block diagram of a modern television transmitting and receiving system.

emitted becomes positively charged, and thus throughout the plate a varying charge pattern is established that corresponds to the light picture focused on the photocathode. The special material from which the target plate is made does not allow the charges to move. Because the target is so thin this charge distribution is established on both sides of the plate. The electrical image stored on the target plate cannot be transmitted as a whole. The individual

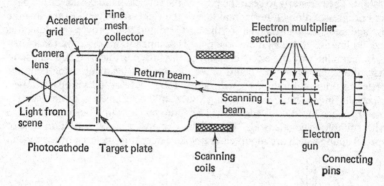

Fig. 136. The image orthicon tube.

picture elements must be scanned in sequence to produce an electrical signal suitable for transmission. This is achieved by an electron-scanning beam formed by an electron gun at the other end of the tube. The formation of this beam is similar to that in a cathode-ray oscilloscope tube. When the scanning beam strikes the target plate a return beam is formed. The number of electrons in the return beam depends upon the position of the scanning beam on the electrical image. Where the beam strikes a positive region of the image, electrons are absorbed and the return beam contains comparatively few electrons. In regions of the plate where little or no positive charge exists almost all of the scanning electrons are reflected into the return beam.

The number of electrons in the return beam, and hence the return current, at any instant is therefore determined by the intensity of the charge at the point on the signal plate that is being scanned at that time. This return current is greatly amplified by the action of the electron multiplier. (This type of multiplier is described in the chapter on photoelectric devices.) It is the special amplifying properties of the electron multiplier that account for the great sensitivity of the image orthicon camera tube. It will be seen that we now have a means whereby the intensity of light and shade in a given picture can be converted into an electrical signal. On making the scanning electron beam travel across the picture, the return beam creates an electrical signal that corresponds to the instantaneous light level at successive positions along the line. Fig. 134 shows the electrical signal that could be produced by scanning one line of the picture.

Scanning and Synchronization

In order to convert an entire frame into electrical form it is necessary to scan the whole scene in a regular and predetermined fashion. The method used is not dissimilar from the process of reading. In order to convey information, ideas are expressed in words and sentences on a printed page. It is not possible to assimilate all of the information on a complete page at one instant. The eye starts to scan the page at the top left-hand corner of the page and progresses along a horizontal line to the right. At the end of the line the eye suddenly jumps to the left of the next line and then progresses along this second line. In this way the whole page is scanned in an organized fashion.

In a similar way the electron beam of the image orthicon tube is made to scan the signal plate. We could move the electron beam by using deflector plates as in a cathode-ray oscilloscope tube, but for television purposes, especially in the large-area viewing tubes in the receiver, a more accurate picture is produced if the scanning is achieved by deflecting the electron beam with suitable magnetic fields. These fields are produced by deflection coils of special shape which are mounted in a yoke external to the neck of the tube in

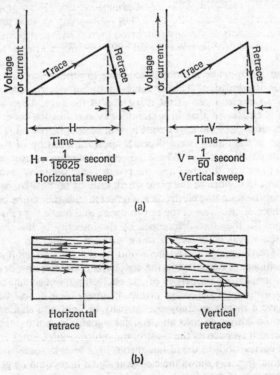

$H = \frac{1}{15625}$ second
Horizontal sweep

$V = \frac{1}{50}$ second
Vertical sweep

(a)

Horizontal retrace

Vertical retrace

(b)

Fig. 137. The scanning process: (a) saw-tooth scanning waveforms for horizontal trace and vertical sweep; (b) progressive horizontal scanning.

the region of the electron gun. The current within the coils is then varied in such a way that the magnetic field rises to a peak value and then suddenly is reduced to its former value. The waveform of the field intensity is shown in Fig. 137. Two sets of coils are required so that deflections can be produced in both the horizontal and vertical directions. The current for producing the sweep trace in the horizontal direction is derived from the sawtooth oscillator and associated amplifier, which together are known as the **line timebase**. In order that the whole rectangular area is covered it is necessary to start each line scan at a progressively lower position. This deflection is produced by another sawtooth generator known as the **frame timebase**. This name arises because one sweep of the frame timebase causes the whole frame to be scanned. To produce the illusion of motion each frame must be scanned at the rapid rate of 25 per second so that eventually when the pictures are reproduced in the receiving equipment, these pictures will merge satisfactorily. Even at this rate the problem of flicker exists. This sensation is particularly noticeable by the outer margins of the retina.

In an attempt to reduce this flicker to negligible proportions a system of interlaced scanning has been adopted. In this system the total number of lines is divided into two groups. The first group consists of every other line, i.e. the even-numbered lines, 312·5 in all. The other group, or field as these groups are known, consists of the odd-numbered lines. The frame is first scanned by the first field and then scanned again by the second field. Fig. 138, in which

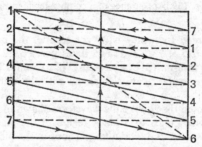

Fig. 138. Interlaced horizontal scanning. First field scans frame on lines 1, 3, 5, etc.; second field scans frame on lines 2, 4, 6, etc.

only 7 lines are shown, illustrates the principle. In the actual television picture there would be 625 lines. We see that on the first run half of the frame information is scanned, but the scanning is distributed over the whole picture area. On the next run the remaining frame information is scanned. Thus although the frame repetition rate is 25 per second, since two fields are shown for each frame, the field repetition rate is 50 per second. This figure is chosen deliberately so as to coincide with the mains frequency. The design of the synchronization circuits is thus simplified.

The need for synchronization has already been mentioned in connection with cathode-ray oscilloscopes. In television it is necessary to have synchronization in order to produce a coherent picture. Having converted the line information into an electrical signal and scanned the field of view in the image orthicon tube, it will be appreciated that to maintain the correct timing of both the horizontal and vertical sweep motions it is necessary to synchronize the line and frame time bases in both the transmitter and receiver. Only by doing this can the receiver reassemble the information into a proper picture. To achieve this synchronization, the transmitter sends out rectangular 'sync' pulses in addition to the picture information. A complete video signal will then have the waveform shown in Fig. 139. Since 625 lines are used and

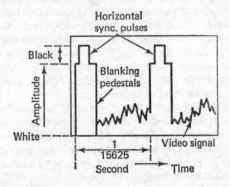

Fig. 139. Appearance of a composite video signal.

each frame scanning frequency is 25 per second, the horizontal time base must operate at 625×25, i.e. 15,625 times per second. A horizontal sync pulse is therefore transmitted every 1/15,625 of a second. This ensures that the start of every line in the receiver coincides with the start of a sweep in the camera tube.

It should be noted that unlike the waveform shown in Fig. 134, the top of the waveform corresponds to a black region of the picture, and the lower amplitudes correspond to the brighter portions of the picture. This is known as **negative modulation** and is used to achieve freedom from noise interference. With **positive modulation** a noise 'spike' produces a brightening of the picture and this would be noticeable; readers may recall the 'snowstorm appearance' of early television sets. With negative modulation, however, a noise voltage temporarily blacks out that portion of the screen being scanned at the time of the noise interference, and this is hardly noticeable by the viewer. Apart from blanking out noise we must also suppress the trace during the flyback period between horizontal scans, otherwise an irritating trace would be produced. This flyback suppression is produced by the blanking pedestal pulse, which must be long enough to cover the required flyback period. By transmitting the horizontal sync pulses during the flyback period such sync pulses cannot interfere with the picture information.

In addition to the horizontal sync pulses, other groups are transmitted between each frame. These are discussed below in connection with receivers.

The Television Receiver

As we have seen, a television system must transmit and reproduce an enormous amount of information. Each line scan contains details of picture elements, and 15,625 lines per second are transmitted. In addition there are the various synchronizing and blanking pulses to be accommodated, apart from a complete F.M. sound channel. It should therefore not be surprising that the bandwidth required for such transmissions is very much larger than that required for sound transmissions alone. The video information requires a 5·5 MHz bandwidth for modern systems, and when combined with the associated sound channel a single television station must use a bandwidth of about 8 MHz. Such a bandwidth cannot be accommodated on the medium-wave broadcast frequencies. For the 625 u.h.f. systems in common use the u.h.f. band is in the range 470 MHz to 854 MHz. The design of the television receiver must take account of the fact that signals having a very much higher frequency than sound radio are involved. Apart from this factor, video and sound amplification and detection is in principle the same as that discussed in the chapter on radio. We shall therefore confine ourselves here to those aspects of electronic systems that are peculiar to television receivers.

The only practical way of producing a television picture at the moment is by using a cathode-ray tube. A great deal of research is being undertaken to produce light emitting diodes of a type from which a mosaic could be produced. Success in this field will mean that flat television sets that could be hung on the wall will be produced. The sets would be little thicker than the picture frames used at present for paintings and photographs. In the meantime we must make do with cathode-ray tubes.

The principles of producing a pencil beam of electrons and drawing pictures on the face of a tube have already been discussed on the chapter on cathode-ray oscilloscopes. The picture size required for entertainment purposes, however, must be much larger than cathode-ray oscillographs to be acceptable to the viewer. For this reason electrostatic deflection of the electron beam has not been found to be satisfactory. In view of the very large deflections required to produce a picture of acceptable size, magnetic deflection must be employed if tubes of reasonable size are to be used. The deflection system of interlaced scanning follows the lines already discussed in connection with the camera tube. The brightness of the picture at any point on the screen is determined by feeding the modulator with the video signal obtained from the line scan. It will be recalled that with an oscilloscope the brightness of the trace is controlled by a knob on the front of the instrument. The knob is connected to a potential divider that controls the voltage to the modulator. The principle of controlling the brightness on a television screen is the same, but this brightness must correspond to the picture information; the video signal is the controlling agency. We see then that a television receiver must incorporate not only video and sound amplifiers and detectors, but also special power supplies to operate the cathode-ray picture tube as well as two relaxation oscillators, viz. the line and frame time bases. In order that these time bases are synchronized with those in the transmitter some means must be found of separating the horizontal line sync pulses from the vertical frame sync pulses. A separation of the video and sound channels must also be effected.

The separation of the sound and video channels is effected at the **video detector stage**. The separation is accomplished by beating together, or heterodyning, the F.M. sound I.F. signal with that of the A.M. video I.F. signal. These I.F. signals are 6 MHz apart. Because of the detector's partially non-linear characteristic, it performs this mixing function automatically. The heterodyning produces a 6 MHz F.M. difference frequency signal which is subsequently processed in the normal way in the sound channel. This signal is separated from the demodulated composite video signal by filter circuits in the output stage of the detector.

A special waveform is transmitted to enable a separation of the horizontal and vertical sync pulses to be made in the receiver. Fig. 140 shows the various stages in diagrammatic form. The top line (Fig. 140(a)) shows the waveform as received. By passing it through a biased diode circuit we obtain the waveform shown in Fig. 140(b). By biasing the diode with a voltage, E, the diode cannot conduct until this voltage has been exceeded. In this way all of the blanking pulses and video information are cut off, leaving only the sync pulses in the sync separator circuits. Each horizontal sync pulse serves to initiate the sweep of each line at the appropriate time. Since we also need to synchronize the frame time base a sync pulse must be supplied for this purpose. The frame repetition rate is much slower than the line time-base rate and so a long vertical sync pulse is required. However, since we have only one carrier wave the vertical and horizontal pulses cannot be radiated simultaneously. To avoid upsetting the synchronous operation of the line timebase during the comparatively long time spent in radiating the vertical sync pulse the latter is serrated (i.e. cut) as shown in Fig. 140(a). We are now in a position to separate the horizontal line sync pulses from the vertical frame sync pulses. The horizontal sync pulses are obtained by using what is called a differentiating circuit. This is a simple *RC* circuit with a short time constant. We see from Fig. 140(c) that irrespective of whether only line or frame pulses are being radiated, the output waveform from the differentiating circuit is always the same. It is this 'spikey' waveform that is used to synchronize the line timebase. It will be appreciated that by serrating the vertical sync pulses we can, with the differentiating circuit, produce the 'spikey' horizontal sync pulses simultaneously with the vertical sync pulse. The frame timebase sync pulse is produced when the serrated sync pulse is passed through a *CR* circuit with a long time constant, the output this time being taken from across the capacitor. Such a circuit is called an **integrating circuit**. It is now easy to separate the short 'spikey' horizontal sync pulses from the much longer vertical sync pulse. The equalizing pulses are provided to smooth out the differences between vertical sync signals for alternate fields. It must be remembered that one vertical sync pulse occurs during the middle of a line for one field but at the end of a line for another (see Fig. 138).

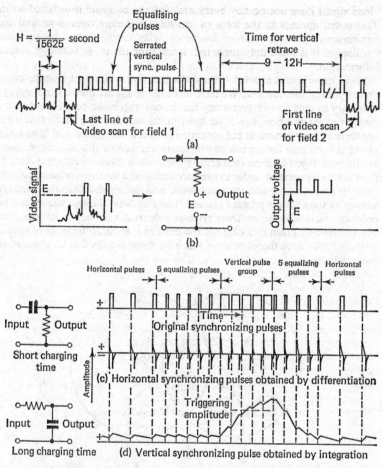

Fig. 140. Separation of horizontal and vertical sync pulses: (a) received waveform; (b) separation of sync pulses; (c) horizontal pulses obtained by differentiation; (d) vertical pulse obtained by integration.

Colour Television

Although the principles involved in the transmission and reception of colour television are similar to those for monochrome (black-and-white) television, the circuits involved are so much more complicated that we cannot discuss them in any detail in this type of book. One of the problems involved is that programmes designed for colour television receivers must also be able to be received by black-and-white receivers. All of the extra information concerned with colour must somehow be compressed into a band-width of 8 MHz. It was discovered as long ago as 1929 that a monochrome video signal does not occupy every cycle of the bandwidth assigned to it. This signal appears in the form of 'clusters' of energy centred around the harmonics of the line scanning frequency, with 'empty spaces' between the harmonics. It is this empty space that is used for the transmission of colour information.

Prism experiments at school reveal the fact that white light is made up of the colours of the rainbow. We have already mentioned that it is not always necessary to give an observer every fact in order to make possible the transmission of information. In colour experiments we have discovered that there are three primary colours: red, green and blue. If red, green and blue lights are combined (say by the use of projectors and colour filters) other colours are formed. Fig. 141 shows the arrangement when three overlapping discs of light fall on a screen. In order to produce white light the three primary colours must be added in the correct proportion. Red, green and blue are therefore known as the **additive primary colours.** Thousands of different shades can be produced merely by adding these primary colours in various proportions. For this reason red, green and blue are the colours that form the basis of colour television. We need therefore to produce electrical signals that correspond to these colours.

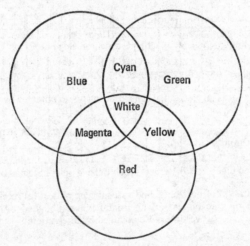

Fig. 141. The colour mixing of the three primary colours red, blue and green.

Brightness, hue and saturation are three terms associated with colour. Brightness or luminance is an indication of the amount of light energy entering the eye. At one time the unit 'candlepower' was used, but nowadays the candela is the unit of luminance. This unit is defined in a special way, being associated with the brightness of molten platinum at a temperature when the platinum is just about to solidify. In monochrome receivers all shades of brightness from black through grey and eventually to white are experienced. The **hue** or **chrominance** of a surface is the actual colour of that surface. Colours of the same family all have the same hue, e.g. light pink, dark pink and red. The **saturation** of a colour is associated with the purity of the colour and is measured in terms of its dilution with white light. A fully saturated colour contains no white light. In order to reproduce a scene in a television studio as a coloured image on a screen, information must be transmitted about the brightness, hue and saturation of that scene.

The Colour Camera

In order to convert a coloured scene into an electrical signal suitable for transmission three individual colour cameras are required. The scene is separated by means of coloured filters into the three primary component colours. Thus we have three secondary scenes, one red, one green and one blue, which when combined will give us the original range of hues. Each secondary scene is viewed with a camera tube. (In some cameras a fourth tube is sometimes used to provide an improved monochrome picture. This tube is not supplied with filters, and is concerned mainly with luminance of the scene.) The colour tubes used in modern cameras are called **plumbicons** because they use target plates that are sandwiches of pure lead monoxide between an *n*-type layer of tin oxide and a *p*-type layer of doped lead monoxide. Such a tube produces high-quality pictures at low light levels. Their small size and light weight make them ideal for incorporation into cameras requiring three or four tubes.

Fig. 142 shows the arrangement of one type of colour television camera. By the use of an ingenious optical system the complete scene is projected into three camera tubes. Each tube has its own colour filter and hence the scene is resolved into its three primary colours.

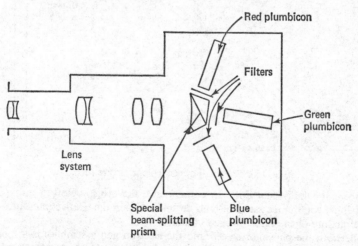

Fig. 142. Simplified diagram of one form of colour television camera.

In the United Kingdom the **PAL system** of transmission is used (PAL stands for Phase Alternate Line). The technicalities of this system are outside the range of this book, but briefly the system was devised to prevent phase errors in the subcarrier waves; such phase errors lead to incorrect hues. The luminance signal in the PAL system utilizes the whole of the bandwidth of the video channel. This signal conveys the brightness information of the transmitted picture and, since this information can be interpreted by monochrome sets, ensures compatability between colour and monochrome systems. The colour information, which refers to hue and saturation, is conveyed by two chrominance signals that modulate both the amplitude and phase of a subcarrier wave.

The major component of interest in the receiver is the colour picture tube. This is a cathode-ray tube which has three electron guns in the neck of the tube, one for each of the three primary colours. The modulators in each of the guns are controlled by the 'red', 'green' and 'blue' signals. The screen consists of a very fine mosaic of phosphor dots. Unlike the tube in an oscilloscope, which has a phosphor that produces only one colour, the television colour tube screen has the mosaic made up with phosphors that emit red, green and blue light. The dots must be distributed in a special way as shown in Fig. 143. At normal viewing distances the acuity of the eye is insufficient to

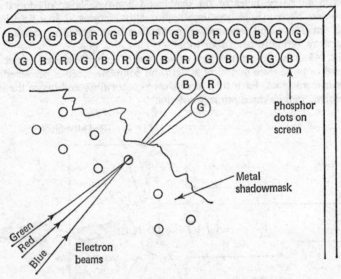

Fig. 143. The shadowmask colour picture tube.

resolve the individual dots. In consequence the eye is deceived into perceiving a continuous range of hues dependent upon the relative intensities in the mixture of the three primary colours.

Between the phosphor screen and the electron gun assemblies is a metal shadowmask, hence this type of tube is known as the **shadow-mask tube**.

This mask has a set of perforated holes (about 450,000 for a 25-inch tube). Each hole is associated with a set of three phosphor dots, one red, one green and one blue. The angles at which the electron beams pass through a given hole determine which dots are energized. The whole of the useful viewing area of the screen is scanned by the three beams which are deflected by the magnetic fields from external deflector coils as in monochrome receivers.

Summary

The television process consists of breaking down a moving picture into a series of still pictures each of which is resolved into lines; the lines themselves give rise to signals which correspond to the intensity of illumination at successive points along the line. This signal is made to modulate a carrier wave suitable for radiation as a television transmission.

The **scanning pattern** (or **raster**) at the picture tube consists of 625 interlaced horizontal lines for each frame, which is broken up into two alternately even- and odd-numbered fields: 25 frames, i.e. 50 fields, are transmitted each second. The persistence of vision is such that the illusion of motion is created.

To keep the transmitter and receiver synchronized, suitable pulses, called **sync pulses**, are transmitted in addition to the picture information.

The **video signal** is transmitted using amplitude-modulated waves. The associated **sound programme** is transmitted using frequency modulation.

The principles of broadcasting colour pictures are similar to those used for black-and-white transmissions. The colour picture information is broken into three components, red, green and blue, and each component is converted into electrical signals by separate **plumbicon tubes**. In the receiver each colour signal controls one of three electron beams which are projected through a **shadowmask** to energize the phosphor dots on the screen. Suitable **phosphors** are used so that the three primary colours can be generated. Because the eye's resolving power is insufficient to resolve the separate dots, an illusion is created of a coloured picture having a range of hues and brightness.

Questions

1. What has been the impact of television on social standards?
2. What difficulties would be encountered if a 1,000-line scanning system were to be used?
3. Explain the reason for using interlaced scanning.

Suggested Further Reading

Jacobowitz, H., *Electronics Made Simple*, Heinemann, 1967.
King, G. J., *Beginners Guide to Television*, Newnes–Butterworth, 1973.
King, G. J., *Beginners Guide to Colour Television*, Newnes–Butterworth, 1973.

12

HIGH-FIDELITY SOUND REPRODUCTION

Characteristics of the Ear

We have seen in the previous chapter how easy it is to deceive the eye when viewing television and cinema programmes. Unfortunately, no such easy deception is possible with the ear. The ear is certainly a very sensitive device, able to detect movements of air molecules that are not much greater than the size of the molecules themselves. The great sensitivity alone, however, is not the only factor involved in hearing. There are many physiological and psychological factors to take into account. Whilst we do not fully comprehend all of these factors, we do have rather complete data on the results of the performance of the ear. We know that the young, healthy ear is able to perceive as sounds those vibrations with frequencies in the range 20 Hz to as much as 20 kHz. To a musician, frequency is perceived as pitch. Each doubling of the frequency of a sound is said to raise the pitch by an octave; thus the ear has a total response over about 10 octaves. Since the highest fundamental pitch of any musical instrument is about 5 kHz, most ears can hear at least the fourth harmonic. The detection of harmonics is a very important property of the ear. This is because the harmonic content of a note determines the special characteristics of the sound, and thus enables us to distinguish between, say, a piano, clarinet and cello each playing the same note. Fig. 144 shows some complex waveforms to illustrate this.

The ear is not equally sensitive to all sound frequencies throughout the range. It is most sensitive to sounds having frequencies in the range of 1 to 4 kHz. The sensitivity falls considerably at low and high frequencies. Usually, the older a person becomes the less is he able to detect the higher frequencies. Many middle-aged people are able to detect sounds only up to frequencies of 10 or 12 kHz. Unfortunately, since the introduction of high-powered discos, the hearing ability of many teenagers and young adults is inferior to that of people of 50 and 60 years of age.

Fig. 145 shows how our ears respond to sounds of different frequencies and intensities. We see, for example, that if we have a tone having an intensity of 10 dB and a frequency of 4 kHz, and compare it with a 20 Hz tone, the low frequency tone must have an intensity of 50 dB to appear to be as loud; i.e. the low-frequency tone must be boosted in intensity by 40 dB to appear to be as loud as the 4 kHz tone.

The intensity of a sound is a measurable quantity in physics and has nothing to do with ears. It is concerned with sound pressures or powers. The loudness of a sound, however, is a subjective quantity and depends not only on the intensity, but also on the sensitivity of the ear, how we are feeling on that particular day, the nature of the sound and other psychological factors. The

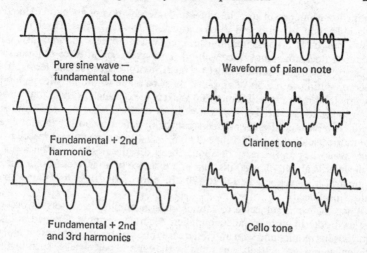

Fig. 144. Complex waveforms. The three on the left show how complex waveforms can be synthesized by adding harmonics to a fundamental sine wave. The remaining complex waveforms are associated with the piano, clarinet and cello.

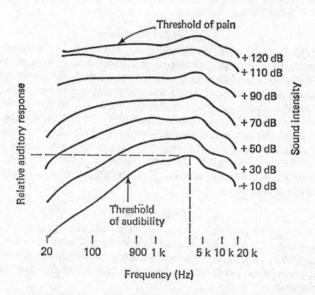

Fig. 145. Curves showing relative auditory response at different sound intensities. Note how for a given intensity the response of the ear (i.e. loudness) falls at high and low frequencies. For example, a 10 dB intensity sound at about 4 kHz sounds as loud as a 50 dB intensity sound at 20 Hz.

distinction is important since loudness and intensity do not go 'hand in hand'. It is important for hi-fi and disco enthusiasts to realize this. Doubling the power output of an amplifier/loudspeaker system doubles the intensity, but the result does not sound twice as loud. To make a sound twice as loud, the intensity, and hence power, must be raised tenfold. In practical terms, if a disco operating with a 50 W amplifier must be made to sound twice as loud, the amplifier rating must become 500 W. It is factors such as these that make it desirable to use decibel scales instead of arithmetic ones. Some idea of loudness levels and decibel numbers can be obtained by noting that orchestral music usually lies in the range of 40 to 100 dB, conversational speech in a quiet room has an intensity of about 20–30 dB, the traffic noise on a busy high street is about 70 or 80 dB, and at a distance of a few feet from a pneumatic drill the level may be 110 to 120 dB. Above 120 dB we find that actual pain is produced.

When the ear is subjected to sounds of large amplitude, distortion occurs just as it does in an overloaded amplifier. A similar result is obtained in that spurious harmonics and sum and difference tones, although not present in the original sound, are actually generated by the ear and perceived as such by the brain. These subjective tones, as they are called, can actually appear as recreated fundamental tones. Hence we experience the sensation of listening to bass tones where none can actually be present. Many cheap loudspeaker and amplifier systems rely on this 'false bass' which the equipment itself is incapable of producing.

High-Fidelity Equipment

High-fidelity equipment is intended to reproduce an original sound programme by electronic and electromechanical means. To earn the title 'high-fidelity' (often abbreviated to hi-fi) the reproduced sound must encompass almost the entire audible frequency range, and the sound field must appear to the listener to be identical, or almost so, to the original. The equipment consists of three component sections: the programme source, the electronic amplifiers and the output sound-producing devices. Programme sources consist of microphones, pick-ups and gramophone discs or records, magnetic heads energized by magnetic tapes, and radio tuners. Sound-producing devices are for almost all practical purposes moving-coil or electrostatic loudspeakers.

Microphones

The microphone may well be considered the most vital component of high-fidelity systems since it is this device that is at the starting point of the whole process. Unless we obtain a satisfactory conversion of acoustical energy into electrical energy the remainder of the system cannot possibly recreate high-quality sound.

The most common microphone in use is the **carbon microphone**, but the common usage is associated with telephone systems, not with high-fidelity systems. This type of microphone is a pressure type in which the sound pressures are made to strike a diaphragm. The vibrating diaphragm is made to vary the pressure exerted on carbon granules, thus altering the state of compression of these granules. The electrical resistance of the granules is a function of their compression, so that if a current is passed through them the magnitude of the current will be altered. The current can then be considered as having a steady component upon which is superimposed a component that varies according to the sound pressure. Such microphones are noisy, and have poor fidelity; they need not be considered for the purposes we have in mind.

The **crystal microphone** actually generates a voltage of its own because it depends upon the piezo-electric effect. We have already seen in the chapter on oscillators that certain crystals generate voltages when mechanical forces are applied. For microphone use the crystal material is usually Rochelle salt and not quartz, as is the case for crystals used in R.F. oscillators. Two wafers, known as a **bimorph element**, are cut and assembled so as to be sensitive to sound pressure waves (Fig. 146(a)). The microphone has a high source impedance that is capacitive (about 1500–2000 pF). Consequently an amplifier which is being driven by a crystal microphone must also have a high impedance so that correct matching is obtained. This impedance may be in the range 1–5 MΩ. The high impedance is a disadvantage in high-fidelity work because it prevents the use of long connecting cables; such cables would be very susceptible to hum and noise, and the self-capacitance of the cable leads to attenuated high-frequency response. The output of a crystal microphone is high, which accounts for its popularity with users of cassette recorders. For this purpose such microphones are inexpensive, but leave much to be desired so far as performance is concerned. The crystal is easily damaged by heat or moisture so the user must take care in this respect. These microphones are usually omnidirectional, that is to say, they detect sounds from all directions with equal facility.

The **dynamic**, or **moving-coil microphone**, has been in use for many years and is still very popular. The main constructional features are shown in Fig. 146(b). This microphone utilizes a coil and permanent magnet; the coil is attached to a diaphragm. The coil consists of many turns of exceedingly thin wire wound on to a former. This former, which is attached to the diaphragm, moves within the pole pieces of a powerful magnet. The e.m.f. produced when the coil moves is proportional to the velocity of the coil vibrations. The electrical impedance of such a microphone is usually about 25 or 30 ohms. A step-up transformer is often used to couple a dynamic microphone to the amplifier. The low impedance enables us to use quite long connecting cables between the microphone and amplifier without suffering a deterioration in performance. It is possible to design transistor preamplifiers that have an input resistance of the order of 30 ohms, so that direct coupling is possible. Such an amplifier must be carefully designed if noise is to be kept down to acceptable proportions. A dynamic microphone can provide a wide frequency response; it will withstand high temperatures and humidity and, being a rugged device, it can tolerate mechanical shock. Unless the highest quality of reproduction is required, the dynamic microphone is suitable for all but the most exacting requirements. The wide and smooth frequency response, good

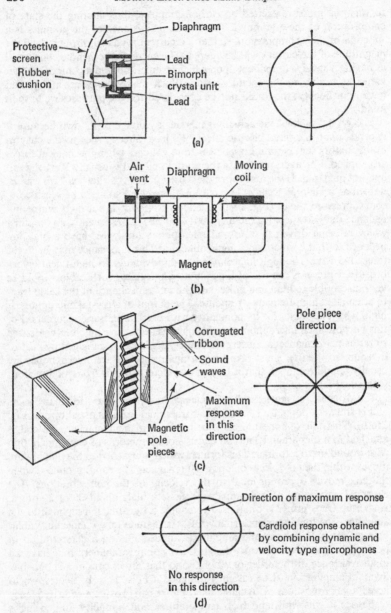

Fig. 146. Constructional features of some commonly used microphones: (a) crystal microphone with omnidirectional response; (b) moving-coil microphone (dynamic type) with almost omnidirectional response; (c) ribbon microphone (velocity type); (d) direction of maximum cardioid response is obtained by combining dynamic and velocity type microphones.

transient response, reliability and comparatively low cost account for the fact that this microphone is used more than any other.

The **ribbon microphone** (Fig. 146(c)) consists of a thin corrugated ribbon of Duralumin foil suspended between two pole pieces of a powerful magnet. The corrugations not only increase the surface area for a given length, but also serve to increase the ribbon's mechanical strength. Sound waves can impinge on both sides of the ribbon, which is so light that the motion of the ribbon corresponds to the movement of the air molecules. An e.m.f. is induced in the ribbon; hence this type of microphone is also a velocity type. The foil is moved by sound waves impinging at right-angles to the surface. As the source moves around the microphone the response diminishes until at 90° to the surface sound waves impinging on the edge produce no response. The response diagram of a ribbon microphone is therefore a figure-of-eight pattern. Such a microphone has a very low impedance and output, hence a transformer is necessary to match into the amplifier. As with the dynamic microphone, long leads to the amplifier are possible without degrading the performance. The ribbon microphone is quite a rugged instrument, but is sensitive to blasts of air. Users must avoid blowing into the microphone to see if the sound system is working.

In many applications the problem of **acoustic feedback is annoying**. Such feedback can arise when we wish to reinforce speech or music by means of a microphone, amplifier, and loudspeaker system all contained within a single hall or room. If the gain of the amplifier is high enough, any acoustic disturbance is detected by the microphone, amplified and reproduced by the loudspeaker system. The sound produced is again detected by the microphone, amplified and reproduced. The original sound disturbance is reinforced to such an extent that the public address system is forced into an oscillation condition thus producing a 'howl' effect. This is an acoustic example of positive feedback. If the loop gain is sufficient a permanent oscillation is established.

One way of avoiding such oscillation is to use a microphone that houses two units, viz. a dynamic (moving-coil) microphone and a velocity (ribbon) type. When the response of the dynamic microphone, which is almost omni-directional, is combined with the figure-of-eight pattern a polar response which is cardioid, i.e. heart-shaped, results (Fig. 146(d)). Polar diagrams are drawn to show how the response varies with the particular direction from which the sound is being received. We can construct a polar diagram by plotting the length of lines, from a fixed point, in various directions. The length of line is proportional to the response in decibels. In order to compare different microphones we usually take the 0 dB line as being a line of convenient length in the direction which the microphone is most sensitive. In all other directions the lines must be shorter, indicating a lower response in those directions. It is convenient to avoid drawing the actual line but merely to indicate where the end of the line remote from the fixed point would be. Fig. 146 gives some examples of typical polar diagrams.

Studio-type capacitor microphones constitute the highest quality of microphone at present available. They have a wide frequency range, a smooth response and excellent transient response. They are often used as reference standards in sound-measuring equipment because they can be so accurately calibrated. They consist essentially of two plates of a capacitor. One plate is fixed and the other is a low-mass movable diaphragm. The capacitance thus varies according to the position of the diaphragm, which in turn depends upon the incidence of sound waves. As the capacitance varies, the potential across it varies since means are taken to maintain the charge constant. It is essential to use an amplifier that is mounted very close to the microphone otherwise cable losses would be very large. Usually the amplifier is physically small and can easily be mounted in the microphone housing.

Disc Records, Pick-ups and Turntables

Following Edison's invention of the phonograph, which used cylinders, Berliner introduced the round flat disc which was the forerunner of all recording discs in use today. The programme material is impressed on the disc by modulating the groove. The recording process involves cutting a spiral groove into soft material with a lathe-like arrangement. The programme material is first recorded on a very high-quality tape recorder, and the tape is then edited before the transfer of the material to a lacquer blank is made. A master stamper is then manufactured from the lacquer record, and this stamper is used to press out mass-produced discs for the market.

Bearing in mind that the disc must turn at a constant rate of rotation, we must recognize the fact that the velocity of the stylus along the groove is greater at the outer edge than it is nearer the centre of the disc. In addition, we have problems concerning the lateral (or sideways) velocity and amplitude of the cutting stylus as it impresses the 'wavy' modulation pattern on the groove. It turns out that for low-frequency modulations cut by magnetic devices very large amplitudes are involved. Unless precautions were taken the cutter would break through the wall of the groove into an adjacent groove. To avoid this type of collapse, signals having a frequency below 1 kHz are progressively reduced. In order to improve the signal-to-noise ratio of the final reproduced programme, signals above 1 kHz are recorded at a progressively increased level. The result is that in order to regain the original frequency response the replay amplifier must boost the low-frequency signals and cut the high-frequency signals. The attenuation of the high-frequency signals not only restores the high-frequency balance, but also attenuates the noise; thus the final signal-to-noise ratio is better than it would otherwise have been.

When reproducing from gramophone records, the first requirement is a turntable, the function of which is to rotate the disc at a constant speed. For nearly all practical purposes a.c. mains operated motors are used. These motors are synchronous types whose speed of rotation is determined by the frequency of the supply. We can thus make use of the good frequency stability of the mains. Alternatively, if the quality of the other reproducing apparatus warrants it, such a motor can be operated from oscillators whose frequency can be adjusted so that the revolution rate of the turntable is precisely correct.

The mechanical engineering of the motors and turntable must be of a very high standard if high-fidelity reproduction is required. There are four main faults from which turntables can suffer: rumble, wow, flutter and speed error. **Rumble** is concerned with the main spindle bearing. If this bearing is not very smooth in action, low-frequency noise is produced. With inexpensive turntables this rumbling noise can be troublesome. With transcription turntables special attention is paid to all aspects of the engineering, including the main bearing; rumble is therefore not a problem with high-grade units.

Wow and **flutter** are due to turntable speed variations. 'Wow' is the term given to low-frequency variations below about 20 Hz. It is often due to eccentricity of the turntable or a poor dynamic balance, which causes a sustained note to vary in frequency and amplitude at a very low frequency. Alternatively, wow may occur because of a faulty idler wheel (Fig. 147). This

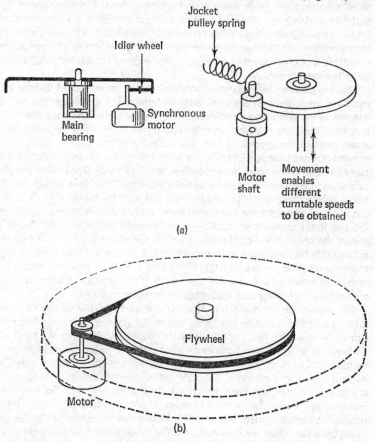

(a)

(b)

Fig. 147. Two types of drive mechanism for gramophone turntables: (a) rim drive motor; (b) belt-driven turntable. The flywheel is attached to the turntable (shown dotted); with belt drives the motor vibrations are isolated from the turntable.

idler wheel in rim-driven turntables makes a mechanical connection between a spindle on the motor and the inner surface of the rim of the turntable. The wheel itself is a rubber or neoprene wheel into which a bearing is inset. Rumble can be caused if this bearing is faulty or inadequately lubricated. The idler wheel's circumference may not be exactly circular, in which case the rim speed cannot be precisely uniform. 'Flutter' is a high-frequency (i.e. above 20 Hz) variation of a sustained note, and is caused by faulty manufacture of the stepped driving cylinder mounted on the motor shaft, or other faulty rotating drive components.

Opinions vary as to the best type of turntable assembly to use. **Belt drives** are advocated by those who wish to isolate the turntable from motor vibration. On the other hand, excellent results can be obtained using a **rim drive**. Constant turntable speed under all conditions, including stylus drag and friction from any cleaning brushes that may be used, implies the use of heavy turntables that have been properly balanced in a dynamic sense. The incorporation of the best engineering principles is essential, and understandably this costs money; consequently the finest turntables will be found to be the most expensive.

The programme information contained in the groove modulations of a record is converted into electrical form with the aid of a **pick-up**. The pick-up consists of an arm and a cartridge containing the stylus and transducer. The arm is so designed that the cartridge can travel freely from the outer to the inner parts of the record. In the vast majority of cases the cartridge is removable from the arm. The **stylus** represents the point of contact between the record disc and the transducer. Modern styli are made from either sapphire or diamond mounted on a suitable cantilever arm. The tip is highly polished into a spherical surface with a radius of 0·001 inch for mono work. The most recent tips, however, have a radius of about 0·0005 inch in order to be suitable for the latest stereo and quadraphonic records. Since such a 'sharp' tip allows the stylus to reach the bottom of the groove on the record (which is undesirable because noise is increased), elliptical tips are now used having a radius across the groove of 0.0007 inch and along the groove about 0·0003 inch. The wide part of the tip prevents bottoming whilst the narrow part allows better tracking with a consequent increase in fidelity.

The **transducer** produces an electrical output by virtue of the vibrations imparted to the stylus as the latter tracks the groove. Any transducer that can convert a vibration into an electrical output could form the basis of a pick-up cartridge. Many have been tried and research proceeds to develop them, but for most practical purposes at present the pick-up cartridges depend upon piezo-electric crystals or magnetic devices. Piezo-electric crystals have already been described in connection with microphones. **Rochelle salt** is frequently used because such high electrical output signals (about 1 V) are obtained; the amplifier design can therefore be simpler than those used for other types of pick-up. The susceptibility of Rochelle salt to moisture and high temperature has led to the development of **ceramic** pick-ups, which work on an electrostriction principle. They are much more tolerant to adverse atmospheric conditions and in addition have a better frequency response than the Rochelle salt type. Unfortunately, a typical electrical output is only 50 to 100 mV. Both types of piezo-electric pick-up are relatively inexpensive.

Magnetic cartridges comprise moving-coil, moving-armature and variable reluctance types. Their construction is shown diagrammatically in Fig. 148.

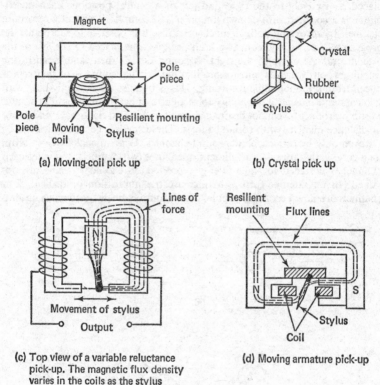

(a) Moving-coil pick up

(b) Crystal pick up

(c) Top view of a variable reluctance pick-up. The magnetic flux density varies in the coils as the stylus moves in the groove

(d) Moving armature pick-up

Fig. 148. Types of pick-up.

In all cases we have a coil associated with a varying magnetic field; either the coil moves within a fixed magnetic field, or the magnetic field is made to change within a fixed coil. In each case an e.m.f. is produced in the coil which corresponds to the vibrations of the stylus. The output from these types of cartridge is quite small, 2–10 mV being typical. The quality of reproduction from well-made magnetic types of pick-up, however, is excellent.

Stereo Pick-ups

The recording of monophonic programmes involves modulating the record groove in a lateral fashion, i.e. in directions along the plane of the disc. This means that on replay a pick-up stylus is made to vibrate horizontally. In stereo recording the groove must carry two sets of programme information corresponding to the left and right hand channels respectively. There is no doubt that stereophonic recording adds considerably to the realism of the reproduced programme.

In this system the original sound programme is 'heard' by two microphones placed respectively to the right and left of a central position. Each microphone is associated with its own amplifier and completely separate recording channel. By recording both channels on a disc it is possible to reproduce the programmes on replay from two loudspeakers, one placed to the left of the listener and the other to the right. A sound field is then established in the listening room which is much more like the original sound than that reproduced from a monophonic recording. This is because we normally hear with two ears and thus have the ability to recognize direction and depth of sound. Mono reproduction cannot distinguish between left and right, back and front, or distance involved with sound sources. These spatial characteristics can be restored only by using separate sound channels. **Conventional stereo** (a term, incidentally, that means three-dimensioned) uses two channels, and when each channel is designed for high-fidelity reproduction excellent results are obtained. In an attempt further to increase the impression of realism, four channels are now becoming popular. With four channels—known as **quadra-**

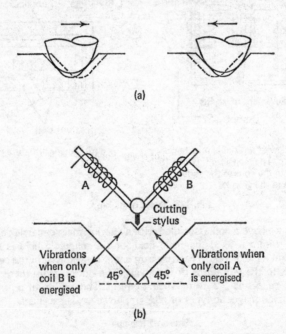

(a)

(b)

Fig. 149. The comparison of groove cutting for (a) monophonic and (b) stereo records. In monophonic recording the stylus is vibrated in the plane of the disc in a horizontal fashion.

phonic stereo, QS for short—we increase the impression of ambience and thus satisfy the ear's instinctive need to hear sound from all directions.

The recording of two-channel stereo programmes on discs involves modulating the two walls of the V-shaped groove independently. The left channel is recorded on the inner wall and the right channel on the outer wall of the groove. Fig. 149 compares the principle of stereo groove cutting with that involved in mono recording. For stereo work the cutting of each wall independently is achieved by energizing either coil A or coil B. In general, both coils are energized simultaneously. The groove walls then are forced into a shape that depends upon the relative magnitudes and phases of the signals in coils A and B.

Fig. 150 shows the principle of operation of a crystal stereo pick-up. The mono pick-up of Fig. 148, shown again in Fig. 150, produces an output when

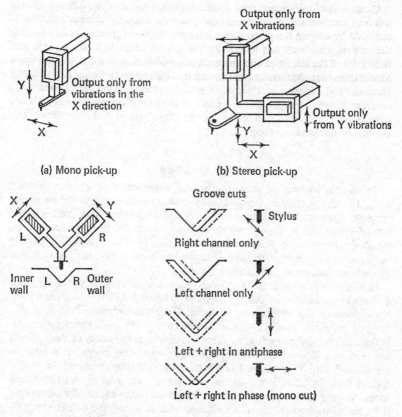

(a) Mono pick-up

Output only from vibrations in the X direction

(b) Stereo pick-up

Output only from X vibrations

Output only from Y vibrations

Groove cuts

Stylus

Right channel only

Left channel only

Left + right in antiphase

Left + right in phase (mono cut)

Inner wall   Outer wall

(c) Stereo pick-up with some basic stylus movements

Fig. 150. Principles of operation of a stereo pick-up.

the crystal is vibrated in the X-direction. No output is produced for movements at right angles to this (i.e. the Y-direction, which is the vertical direction when used as for purely monophonic purposes). If we now couple two mono pick-ups mechanically (Fig. 150(b)) and arrange that the crystals and housings are mutually perpendicular we have formed a stereo pick-up. Such an arrangement is mounted in a headshell in a position shown in Fig. 150(c). Movements in the Y-direction caused by modulations on the outer wall of the groove yield an output only from crystal R, and hence the electrical signals for the right-hand channel are produced. Movements in the X-direction caused by modulations on the inner wall of the groove ideally produce no output from crystal R. They do, however, energize crystal L and hence give us a signal for the left-hand channel. Magnetic pick-ups yield the right- and left-hand signals in a similar way except that electromagnetic principles are involved in the production of the signal voltages.

Quadraphonic discs are now appearing upon the market. Since a groove has only two walls, each wall must carry two channels of information. This is achieved by coding the programme information electronically before cutting the record. One wall has a modulation that is equivalent to the sum and difference of the left front and left back microphones. The other wall carries a modulation that is equivalent to the sum and difference of the signals obtained from the right front and right back microphones. During replay electronic circuitry is used to separate the signals of the four channels. The signals are subsequently amplified and fed to four loudspeakers placed near to the four corners of the listening room.

Tape Recorders

An alternative way of recording sound programmes is to use magnetic tape. The development of high-quality cassette recorders has led to an increase in the popularity of tape recorders and a corresponding decline in the use of discs. The ease with which the user can make his own recordings, especially from radio programmes, make this type of sound reproducing equipment increasingly attractive. (The reader should ensure that he does not infringe any copyright regulations in force.)

In disc recording we deform a soft groove in a plastic disc in such a way that a physical modulation of the groove shape is produced. The modulation must correspond to the programme material. Once this principle had been discovered, research workers sought to record sound programmes by modulating the physical properties of other materials. Two results of their efforts are now in wide use. By varying the intensity of the image on a strip of photograph film it is possible to reproduce the sound by detecting the modulations with a photoelectric cell. This is the system used in cinematograph films. As long ago as 1898, Poulsen in Denmark explored the idea of varying the state of magnetization of a thin steel wire at successive points along its length. Such a system could be made to reproduce sound by detecting the magnetic modulations. The wire was passed at constant speed past a pick-up head which consisted of a coil of wire wrapped around a ferromagnetic pole piece. The varying magnetic modulations then induced voltages in the coil that corresponded to the sound programme. Such a system did not produce recordings of high quality and subsequently fell into disfavour.

During the Second World War the Germans had developed a plastic tape coated with a fine layer of extremely tiny iron oxide particles of uniform size. The fidelity of recordings produced by such a tape was so remarkable that work on steel wire was dropped. Subsequent research to produce tapes coated with improved ferromagnetic powders has led to the production of very high-quality recordings. The fidelity of tape recordings is so high that even disc records are produced by first recording the programme as a high-fidelity tape recording.

Recording

The myriads of small magnetic particles of iron oxide in the coating of a tape act like tiny magnets. In the neutral or unrecorded state, these particles have a random orientation and hence there is a cancellation of their combined magnetic fields. We are then said to be using virgin tape. When the tape is passed by a recording head the individual particles are subjected to a magnetizing force that depends upon the strength of the instantaneous magnetic field produced by an audio signal fed into the head. The head itself consists of a small, carefully manufactured ring magnet. A gap in the magnet allows the magnetic field to spread out and thus influence the tape which is passing very close by.

Fig. 151 shows the principle. The recording head consists of a laminated core and the gaps are filled with a non-magnetic shim. The front gap width must be very narrow if the high frequency response is to be satisfactory. The length of that portion of the tape that is magnetized in order to store one complete cycle of the signal waveform depends upon the speed at which the tape is passing the head and the frequency of the signal. In order to be economical in the use of tape, speeds of $7\frac{1}{2}$, $3\frac{3}{4}$ and $1\frac{5}{8}$ inches per second are commonly used. If the frequency of operation is 10 kHz, for example, then 10,000 cycles must be fitted into $7\frac{1}{2}$, $3\frac{3}{4}$ or $1\frac{5}{8}$ inches. If we take $3\frac{3}{4}$ inches as an example, each cycle will occupy $3.75 \div 10,000$, i.e. 0.000375 inch. This is only 0.375 thousandths of an inch. Since the head gap must be small compared with the length of tape occupying one cycle, it will be seen that a high degree of engineering skill is required in the manufacture of the head. Since there is a practical limit to the width of the gap that can be satisfactorily manufactured, we must expect a fall in the frequency response at the upper audio frequencies. The slower the tape speed the more acute is this problem. In an attempt to make good the head losses at the upper frequencies, considerable boosting of the higher frequencies is required during the recording process. We will refer to this again after discussing the magnetic recording process.

Materials vary in their ability to be magnetized. We find that some materials are very easy to magnetize, but that they do not retain their magnetism when the inducing field is removed. This ability to retain magnetism is called

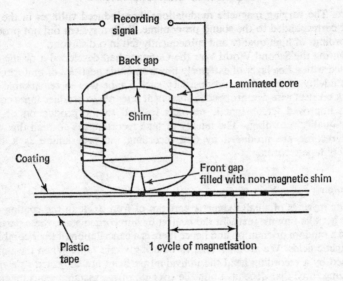

Fig. 151. Principle of magnetic recording.

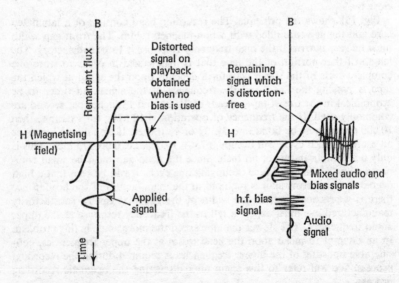

Fig. 152. Principle of using high-frequency bias to reduce distortion. The high-frequency bias signal and audio signal are added together. It should be noted that simple addition produces the waveform shown; no modulation of the bias signal is involved.

coercivity. If we were to measure the remaining magnetism (called **remanent induction**) after magnetic fields of varying strength were applied we would discover a disappointing result. The remanent induction is not proportional to the inducing field. That is to say, if, after removing a magnetic field, we measured the magnetism remaining in a given substance, we would find that had we doubled the magnetic field the remaining magnetism would not have been doubled. The graph showing the relationship between the remanent induction (B) and the magnetizing field (H) is not a straight line; the graph is shown in Fig. 152. This means that if a signal is applied to the recording head, the remanent flux on the tape becomes distorted. Since it is the remanent flux that induces voltages in a similar (or the same) head during playback, a distorted sound reproduction is produced. It has been discovered, however, that if the recording signal is mixed with a supersonic sinusoidal signal we can shift the operation from the non-linear portion of the transfer character-istic to a linear region. Fig. 152 shows diagrammatically how this is done. The signal waveform has in effect been shifted, or biased, so as to operate on the linear portion. The supersonic signal that produces this effect is called the **bias signal**; its frequency must be very much higher than the highest audio frequency being recorded in order to prevent beats. Frequencies commonly used lie in the region 40 kHz to 100 kHz. Since such a high frequency cannot be recorded with the type of head used with audio tape recorders, only the distortion-free recorded signal remains.

A complete reel type of tape recorder is shown diagrammatically in Fig. 153. It will be noticed that the tape first passes over an **erase head**. This is a head with a gap very much wider than a recording or playback head. It is fed with a high-frequency signal; the oscillator that supplies the high-frequency bias is used to drive a power stage which in turn energizes the head. The function of the erase head is to saturate the tape and thus obliterate any signal previously recorded on the tape. In Fig. 153 a **record head** and a **playback head** are shown. Often, for reasons of economy, only a single head is used. When recording, the head is switched to the output of the recording amplifier; on replay the head is switched to the input of the playback amplifier. As with gramophone turntables, the engineering of the deck must be of a high order if the tape is to be drawn over the heads at a constant speed. Defects in any of the rotating parts can lead to 'wow' and 'flutter'. Facilities are provided for switching the reels to driving forces that enable a fast rewind or fast forward movement of the tape to be obtained.

Very compact tape recorders are now made in which the two reels are incorporated in a box called a **cassette**. Cassette recorders are very convenient to use since the cassette is merely plugged into the deck. Such recorders were invented to avoid the inconvenience of positioning reels and threading tape through the deck system; these recorders therefore rival disc machines in convenience of use. Their principle of operation is identical to the open-reel type.

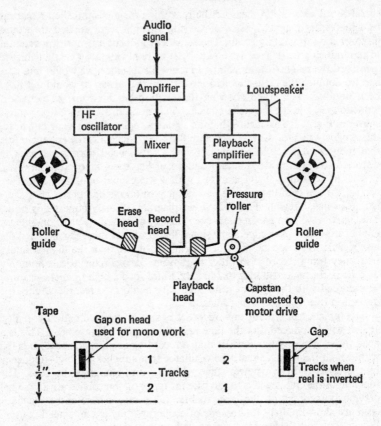

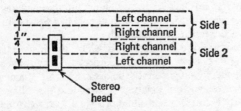

Fig. 153. Main components of a tape recorder. Often a single head serves the function of recording and playback by the use of a suitable switching circuit. The lower diagrams show how the tape area is used for mono and stereo recordings.

High-Fidelity Amplifiers

The principles of operation of voltage and power amplifiers have already been discussed in previous chapters. Readers may care to refresh their memories by looking again at the closing paragraphs of Chapter 4 dealing with audio voltage amplifiers. We shall concentrate here on those aspects of preamplifier design that are particularly relevant to the reproduction from gramophone records and tape recordings.

So far as disc records are concerned, we have already seen the need to attenuate the low frequencies and boost the high frequencies during the recording process. The former precaution avoids excessive groove deformations at low frequencies with a possibility of breaking through from one groove to the next; the latter improves the signal-to-noise ratio, since on replay we reduce the boosted high frequencies to their normal level and in doing so reduce the noise due to record and amplifier hiss. In order to ensure that the correct tonal balance is obtained when playing records manufactured by different companies throughout the world, these companies have agreed with an International Electrical Commission to standardize the amount of bass cut and treble boost used. This results in a standardization of the playback characteristics of amplifiers used for disc reproduction. This standard is known as the IEC (International Electrotechnical Commission) standard, formerly the RIAA standard (RIAA stands for Recording Industry Association of America).

The two-transistor type of equalization preamplifier shown in Fig. 49 suffers from certain disadvantages. In the main, the open-loop gain is insufficient to give a final closed-loop gain at different frequencies so that overloading on programme peaks is avoided. One way out of this difficulty is to use an **integrated-circuit (i.c.) operational amplifier**. Considerable improvements have been made recently in the production of good low-noise i.c. amplifiers for audio purposes. The SN76131 of Texas Instruments and the SGS-Ates types TBA231 are good examples. Fig. 154 shows the integrated circuit equivalent of Fig. 49 and uses a Motorola MC1303. The correct playback preamplifier response is also given, showing the boosting and attenuation required at low and high frequencies respectively. If crystal pick-ups are used a complicated compensation circuit is unnecessary since the characteristic of such a pick-up has a rising response at low frequencies and a falling response at high frequencies. The feedback path may then be a single resistor calculated to give the required gain. The pick-up must, however, be connected to a high impedance, but this is no problem since a 1 or 2 MΩ volume control variable resistor can easily be used at the input of the preamplifier.

Tape Recorder Amplifiers

The design of amplifiers for use with a tape deck is strongly influenced by the characteristics of the tape and electromagnetic recording/playback head. During playback the magnetized tape is passed over the face of the playback head and is in close contact with the gap shim. The varying magnetic state of the tape induces voltage in the coils of the head.

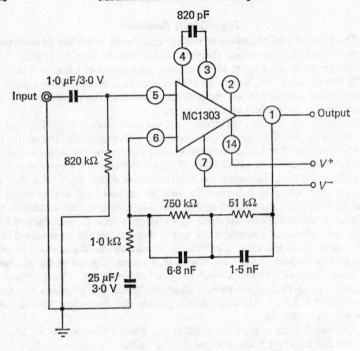

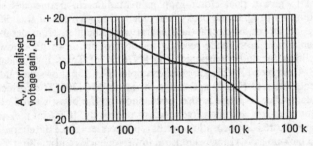

f, frequency, Hz

Typical performance characteristics
pick-up amplifier equilised to RIAA characteristics

Voltage gain	: 34 dB (50) @ 1·0 kHz
Input overload point	: 100 mV rms @ 1·0 kHz
Output voltage swing	: 5·0 V rms @ 1·0 kHz a 0·1% THD.
Output noise level	: Better than 70 dB below 10 mV Phono input (input shorted).

Fig. 154. Circuit for use as a magnetic pick-up preamplifier equalized to IEC (RIAA) standards. The graph shows the response of the preamplifier when the frequency-dependent feedback network shown in the circuit diagram is used.

Consider the position when we have recorded along the tape a signal of rising frequency. If we arrange that the strength of the magnetic elements induced on the tape is the same at all frequencies, then the voltage produced by the replay head will not be the same at all frequencies. This is because the induced e.m.f. is proportional to the *rate of change* of magnetic flux. Hence even if the flux is the same at all frequencies, the rate of change must rise as we increase the frequency. Every time we double the frequency the rate of change must double and hence the output voltage from the head must double. This means that the output voltage from the head rises with frequency at the rate of 6 dB per octave.

At some frequency determined by the construction of the head, losses associated with the head become severe so that the input voltage does not continue to rise, but falls severely as shown in Fig. 155. At first sight it would seem that the playback characteristic should be a mirror image of the graph shown in Fig. 155, i.e. the playback amplifier should boost the bass and treble frequencies. There is a limit to the amount of bass boost we can apply,

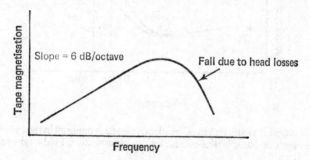

Fig. 155. Playback head output for a constant-current recording.

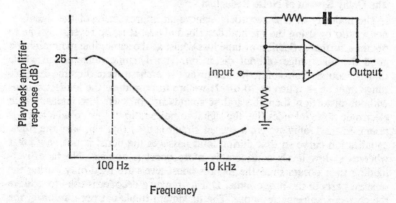

Fig. 156. Practical playback amplifier response (see text). A series *CR* circuit is used in the feedback of an operational amplifier. As the frequency of operation falls the feedback impedance rises; hence the gain of the amplifier rises.

because the 50 Hz hum content will be increased in an undesirable way; we therefore limit the amount of bass boost. An enhancement of the treble frequencies is also undesirable because of the increased hiss that would be produced. The way out of the difficulty is to enhance the bass frequencies during playback and to increase considerably the treble content during the recording process. Not only do we wish to make up for head losses, but we want further enhancement so that on replay a useful improvement in the signal-to-noise ratio can be obtained. It will be recalled that this procedure is also adopted when recording discs. Figs. 156 and 157 show typical playback and recording characteristics.

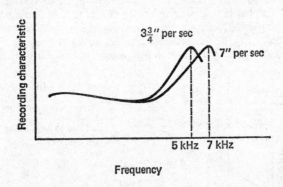

Fig. 157. Recording characteristic. Considerable boost is required in the treble region. The amount of boost is determined, among other factors, by the tape speed.

The Dolby System of Noise Reduction

The majority of tape recorders achieve an improvement of the signal-to-noise ratio by using the method described above. It is, however, possible to obtain a further reduction of tape hiss whilst at the same time maintaining a good dynamic range without distortion. The dynamic range is the whole range of sounds from soft to loud. During live orchestral concerts the dynamic range may be as much as 70 dB. If we are to reproduce the loudest sounds without distortion then the softest sounds are of such low intensity that electronic noise intrudes (i.e. the signal-to-noise ratio at low volume levels is poor). In the Dolby system a special circuit is used that can vary the treble equalization curve so that during loud passages the boost is not very great whereas at low levels considerably more boost is applied. No tape overloading then occurs since the highest boost levels are used only during the quietest parts of the programme. During replay a decoder is used to achieve the correct frequency response. The maximum treble cut occurs during the replaying of quiet passages. This cut also reduces noise and hiss, and consequently almost noiseless reproduction is achieved. The system is most effective at the slower tape speeds, and hence is of greatest value in cassette recorders.

Tone Controls

Even though the initial stages of a preamplifier are designed to match the input transducer, and provide equalization, it is still desirable that the overall frequency response of the system is to a certain extent under the control of the user. This enables allowances to be made for variations in loudspeaker performance and for the different acoustics encountered in a variety of room shapes and furnishings. We need to be able to boost or cut the bass and treble frequencies at will. This is achieved by having frequency-dependent impedances in a suitable circuit arrangement associated with a feedback amplifier.

Fig. 158 shows a typical tone control circuit due to Baxandall. This type of circuit is in almost universal use. Once again we must remember that with a

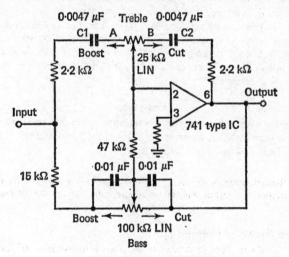

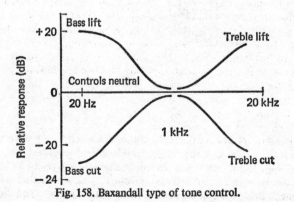

Fig. 158. Baxandall type of tone control.

feedback amplifier the gain depends upon the ratio of feedback impedance to input impedance. Reducing the feedback impedance reduces the gain provided the input impedance remains the same or is increased. If we increase the feedback impedance or decrease the input impedance the gain is increased. When these changes of impedance depend upon the frequency of operation (because the impedances themselves include capacitors or inductors which have reactances that are frequency dependent), it is possible to alter the frequency response of the overall amplifier. By making the feedback and input impedances variable we can construct variable filters and tone controls. In our tone control, consider for example what happens when the slider of the treble potential divider moves towards the point B. The input impedance is increased; this impedance is then the 0·0047 μF capacitor, C1, in series with 25 kΩ when the slider is at B. The feedback impedance is reduced to only the capacitor C2. The higher the frequency of operation the less is the reactance and hence impedance of the feedback path. The gain is thus reduced by an increasing amount as the frequency of operation increases; we have achieved treble cut. By considering how the feedback and input impedances vary with different positions of the controls the reader should be able to deduce how tone control is achieved.

Loudspeakers

The output from the preamplifier equalization sections and tone controls is fed to a power amplifier, which in turn drives the loudspeaker. Power amplifiers have been discussed in Chapter 6.

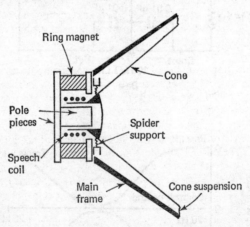

Fig. 159. Constructional features of a moving-coil loudspeaker.

Loudspeakers are devices that recreate sounds when fed by the appropriate electrical signals. The vast majority of loudspeakers used today are **moving-coil** types. Fig. 159 shows the main constructional features of a moving-coil loudspeaker. When a current is passed through a coil of wire that is immersed in a magnetic field, the coil experiences a mechanical force. The direction of the force depends upon the relative directions of the magnetic field and electric current. If the magnetic field is fixed, changes in direction of current produce changes in direction of the mechanical force. Oscillating currents therefore produce oscillating forces, and if the coil is free to move then it will vibrate. In doing so, sound pressure waves will be produced. If the current in the coil varies according to a sound programme then that programme will be reproduced by the moving coil. By considering these facts we can deduce the main constructional features that must be incorporated into a loudspeaker unit. The coil must move in a uniform magnetic field; this is achieved by paying attention to the magnet design. If the field were not uniform then equal changes of current would not produce equal changes of sound; distortion would result. Also intermodulation distortion would result if a large low-frequency signal were applied simultaneously with a small high-frequency signal in a non-uniform field. This is because the high-frequency movements would vary depending upon the instantaneous value of the low-frequency signal. At times when the low-frequency signal is at its maximum value the coil would be moved into a region of weaker field and the high-frequency vibrations would be reduced compared with those produced when the coil is in its central position in the strongest parts of the field. The amplitude of the high-frequency sound would therefore be modulated by the low-frequency signal.

The coil alone could not produce very loud sounds when it vibrates. In order to increase the sound level the coil is wound on a former which is attached to a large cone. Vibrations of the coil make the cone move in a piston-like fashion and thus pressure waves of large amplitude are established. It is a matter of concern to the loudspeaker designer to obtain the correct acoustic coupling between the vibrating coil and the mass of air in the room.

Loudspeaker Enclosures

The coupling can be improved by enclosing the loudspeaker unit in a cabinet of suitable design. For good bass response the loudspeaker should be mounted on a flat board called a baffle. Since the vibrating cone produces sound to the rear as well as the forward direction we wish to avoid a cancellation of the two sounds. The acoustic path length from the front of the cone to the back should be very long compared with the wavelength of the lowest tone being reproduced. Since the velocity of sound in air is about 330 metres per second, and velocity = wavelength × frequency, we see that an organ or bass guitar note of, say, 30 Hz has a wavelength of 11 metres (or about 35 feet). There are not many rooms that could accommodate a flat board some 35 feet square. The practical solution is to use an infinite baffle cabinet. Such a cabinet is in effect a closed box made from dense wood. Chipboard of a type used for flooring is satisfactory; it is dense and about $\frac{3}{4}$ inch in thickness.

The size of the cabinet is dictated by the type of loudspeaker unit used. Manufacturers of these units will always be happy to supply drawings showing how cabinets of suitable dimensions should be made. Fig. 160 shows a typical arrangement. Many other more complicated constructions are possible and the reader is referred to books on loudspeakers for the details. The

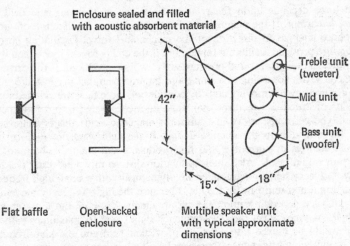

Fig. 160. Various forms of loudspeaker housings.

objects to be achieved in the design of cabinets are to obtain efficient coupling to the mass of air in the room and to avoid resonance and other effects that 'colour' the sound, thus reducing the fidelity of the reproduction.

Multiple Speaker Assemblies

It is not possible to manufacture a single loudspeaker that can handle the whole audio-frequency range at the powers involved. We have already seen how intermodulation distortion can arise when using a single speaker unit. These difficulties can be overcome by using several loudspeakers in one cabinet enclosure, and arranging that each loudspeaker handles only a portion of the frequency spectrum. In jargon terms the large-coned speaker that handles the bass frequencies is called the **'woofer'** and the small speaker that handles only the upper frequencies is called a **'tweeter'**.

It is necessary to interpose between the power amplifier and the loud-speakers a filter circuit called a **crossover network**. Such a network divides the complete audio programme from the amplifier into suitable frequency ranges and feeds these parts of the signal to the appropriate loudspeaker. Fig. 161 shows some typical crossover networks.

F.M. Stereo Radio

Stereo reproduction involves the use of two completely separate channels commencing with two microphones and finishing with two amplifiers and a pair of loudspeakers. It may at first sight seem necessary for a stereo radio transmission and reception system to use two transmitters and two receivers. Fortunately this is not necessary because we can use an ingenious method of **multiplex transmission.**

We have already discussed the way in which a single groove on a gramo-

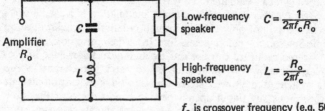

$$C = \frac{1}{2\pi f_c R_o}$$

$$L = \frac{R_o}{2\pi f_c}$$

f_c is crossover frequency (e.g. 500 Hz)
R_o is speaker impedance (e.g. 8 or 16 Ω)

(a) Quarter-section network

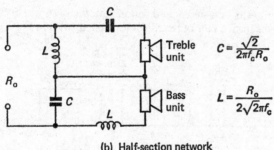

$$C = \frac{\sqrt{2}}{2\pi f_c R_o}$$

$$L = \frac{R_o}{2\sqrt{2}\pi f_c}$$

(b) Half-section network

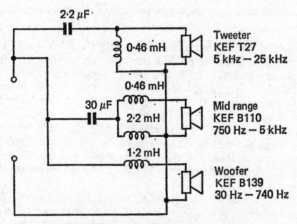

(c) Comprehensive crossover network

Fig. 161. Examples of frequency divider circuits for multiple-speaker systems (note that capacitors should not be electrolytic types). (a) Quarter-section crossover network for two-speaker systems. Filter characteristics rise and fall at only 6 dB per octave. (b) Half-section crossover network. Filter characteristics rise and fall at 12 dB per octave. (c) Crossover network of Heathkit AS-9530 loudspeaker system.

phone disc record can be modulated in a way that carries four-channel stereo information. We have also seen how a complex modulation of a carrier wave can be made to convey not only video information but also synchronization and colour information for colour television. It is, therefore, not surprising that for F.M. stereo radio we can use our technological knowledge in producing complex carrier waves suitably coded to carry two-channel stereo messages. The system used allows a stereo signal to be broadcast on a single v.h.f. channel without any significant increase in bandwidth over that used for monaural transmissions; such a system gives stereo signals at least as good as those obtained from the best stereo records, and at the same time is mono compatible, i.e. mono receivers can be used to receive the programme material.

The various components of a stereo modulation pattern are shown in Fig. 162. The stereo signals from a pair of microphones are amplified in the usual

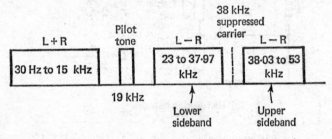

Fig. 162. Various modulation components of a frequency-modulated stereo transmission.

way at the transmitter, but before transmission the right and left signals are processed by a matrixing circuit to produce one signal corresponding to the sum of the two stereo signals (i.e. L+R) and another signal corresponding to the difference of the stereo signals (i.e. L−R). These two new signals are then made to modulate carrier waves. The L+R signal modulates the carrier in the normal way. To convey the information contained in the second new signal a very stable 19 kHz pilot tone is generated which is used as additional modulation and when detected serves to synchronize the receiver. The 19 kHz tone is frequency doubled and this 38 kHz signal provides an R.F. carrier for the L−R signal. In practice the 38 kHz carrier is suppressed and only the sidebands are transmitted.

A mono receiver detects only the L+R signal; since this signal combines the left and right information it is monophonic. In stereo receivers two types of decoder are possible, viz. the sum and difference decoder or the switching decoder. In the sum and difference type the two new signals are first separated by filters. To obtain the left-hand channel the two signals are added, thus (L+R) + (L−R) = 2L. To obtain the right-hand channel the two signals are subtracted, thus (L+R) − (L−R) = 2R. The left and right components are separately amplified and reproduced by a stereo amplifier and a pair of speakers. In the second method of decoding a switching decoder is used.

The 19 kHz pilot tone is separated from the rest of the information and is used to produce a 38 kHz subcarrier. This subcarrier is then used to switch two demodulators that have the (L+R), (L−R) and 19 kHz pilot tone fed to the input of each of them. Each demodulator then produces a left and right channel respectively. Block diagrams of the systems are shown in Fig. 163.

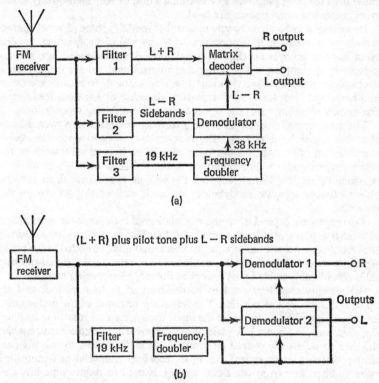

(a)

(b)

Fig. 163. Demodulators for stereo transmissions. (a) Sum and difference decoder. Filter 1 passes only audio signals up to 15 kHz; filter 2 passes frequencies only above 23 kHz; filter 3 passes only 19 kHz pilot tone. (b) Switching decoder.

For the best stereophonic listening the speakers should be placed about 6 to 10 feet apart along the shorter side of the room. If the room has a shorter side of length x feet then the speaker separation should be $0.6x$ feet; for example, if speakers are arranged along a wall 10 feet long they should be placed 6 feet apart.

Summary

The ear is the device that converts sound pressure waves into nerve signals that are interpreted by the brain. Several psychological as well as physiological factors are involved in perceiving sound. The ear can detect frequencies in the range 20 Hz to 20 kHz, but this range reduces with the age of the listener.

High-fidelity equipment is intended to reproduce an original sound pro-gramme by electronic and electromechanical means. The reproduction must appear to the listener to be almost indistinguishable from the original sound.

Microphones are transducers that convert sound-pressure waves into a corresponding electrical signal. The dynamic and crystal microphones are those used for most purposes. For scientific work, or very high-quality studio work, capacitor microphones are used.

Recordings can be made by any means that modifies some physical charac-teristic of a medium in accordance with the sound programme. In **disc** recording the grooves on the disc are deformed or modulated into a wavy pattern by the cutting stylus. The pattern corresponds to the sound signal. Precautions must be taken to avoid breaking through from one groove to another when recording low frequencies. Boosting of the high frequencies during recording improves the signal-to-noise ratio during reproduction.

The programme information in the groove modulations is recovered as an electrical signal by a **pick-up**. The pick-up may depend upon the piezo-electric effect of crystals, the electrostriction effect in certain ceramics or the electromagnetic effect. For mono work the stylus is made to move in a horizontal direction. For stereo work the pick-up is constructed so as to be able to follow separate modulations on each wall of the groove on the record.

Tape recorders depend upon the modulation of the state of magnetization of a ferromagnetic powder deposited as a thin film on a plastic tape. During the recording process a high-frequency bias signal must be added to the audio signal to reduce distortion to a minimum. To improve the signal-to-noise ratio considerable treble boost is applied during the recording process.

High-fidelity **amplifiers** need to be designed so as to introduce as little noise and distortion as possible. The frequency response of the preamplifier must be 'tailored' so as to match the input transducer and recording medium (i.e. disc, tape, radio, etc.). A variable filter, known as a tone control, enables the user to adjust the overall response of his audio system to suit his par-ticular listening environment. In tape recorders a special self-controlled variable filter, known as the **Dolby system**, is used to reduce tape hiss and noise.

Loudspeakers, driven by power amplifiers, convert the electrical signals into audio sound programmes. To work efficiently these loudspeakers must be housed in a suitably designed cabinet. For hi-fi purposes the signals from the power amplifier must be divided with respect to frequency by a **crossover network** into two or three ranges. Each range is then fed to an appropriate speaker.

High-fidelity **stereo broadcasting** systems use frequency modulation. A single transmitter and carrier wave system can be used for stereo work by processing the left- and right-hand channel information electronically in a special way. **Decoders** are then needed in the receiver to recover the left- and right-hand channel information.

Questions

1. Why is high-fidelity sound reproduction so popular?

2. Which type of microphone would you select for use in connection with a cassette recorder?

3. Are crystal or magnetic pick-ups to be preferred for hi-fi systems?

4. Why is it necessary to use crossover networks in connection with loudspeaker systems?

5. Why must the initial stages of an audio amplifier be correctly matched to the transducer?

6. What are the factors that determine the choice of discs or tape as the recording medium?

Suggested Further Reading

Babani, B. B., *2nd Book of Hi-Fi Loudspeaker Enclosures*, Bernards, 1974.

Wireless World Publication *High Fidelity Designs*, IPC Electrical-Electronic Press, 1974

Regular reading of *Wireless World*, *Practical Electronics* and *Hi-Fi Answers* is a good way of keeping up to date with the latest developments in this field.

13

INTRODUCTION TO DIGITAL CIRCUITS

The progress of modern industry depends to a large extent on our ability to introduce an increasing amount of automatic control into industrial and commercial processes. The manufacture and assembly of machine parts, the precision drilling and cutting of raw materials, the automatic sequencing of sorting, weighing, checking and storing of raw materials, and the automatic control of chemical and other processes are all profitable fields for the introduction of automatic techniques. Such techniques could not be developed without the aid of electronics and particularly those branches that incorporate digital circuits. To an increasing extent the routine and often mundane work once performed by human beings is now being undertaken by machines that are controlled by electronic means. The digital computer is a good example of a sequencing machine that performs commercial as well as scientific calculations. Banking and accountancy are two fields in which digital computers are being used to an increasing extent. Police records and motorcar registration records are now being kept in computer banks.

In a more spectacular field, journeys to the moon and other planets would be impossible without the aid of digital computers. Even launching a weather or communications satellite needs computer-controlled devices to enable the rocket to go into orbit. In the home, logic circuits are being used in such apparatus as digital clocks, stereo decoders and electronic musical instruments.

With all industrial control apparatus the main consideration is concerned with the problem of **reliability**. For this reason control systems are based on two states—namely, 'on' and 'off'. The system has to detect only when a signal is 'on' or 'off', and this it can do with great reliability. These two-state, or **binary**, systems form the basis of all control and computing equipment. We could use electromagnetic switches, i.e. relays, to give us an 'on' or 'off' function. In fact relays are still used extensively in many industries, particularly in telephone exchanges, but such switches are not fast enough or reliable enough for use in computer type equipment.

Boolean Algebra

Boolean algebra is the mathematical language of digital systems. This form of algebra was devised by George Boole in the 1840s to help solve problems associated with logic and philosophy. It is concerned with the relationships between **classes** or **sets**. Each class or set is represented by a letter of the alphabet. Although in this form of algebra the letters and signs are borrowed from ordinary algebra, we must rid our minds entirely of all ordinary mathematical meanings. The letter or symbol used to represent a class tells us nothing about the size or magnitude of the class. We make logical statements about any class and these statements must be **true** or **false**; no half-truths or shades of meaning are allowed. This corresponds to our on/off arrangement. A switch is either on or off; we cannot have any intermediate state such as 'half-on' or 'nearly-off'.

When a statement is true it is said to have a logical truth value of 1. Thus if we consider the statement 'rain is wet', we could represent that statement by the letter A, i.e. A = 'rain is wet'. Since the statement is true we say A = 1. It should be emphasized that here the '1' is not 1 in the sense of being half of two or a third of three; A = 1 merely means 'It is true to say that rain is wet.' If B = 'Snow is black', then since this statement is not true we write the Boolean equation B = 0. '0' and '1' are the only allowed truth values; hence we are dealing with a binary system.

We can combine statements and ask, 'Is it true or false to say that rain is wet AND snow is black?'. For the combined statement to be true both statements must be true. Thus since A = 1 but B = 0, the combined statement A.B = 0 (i.e. it is not true to say that rain is wet AND snow is black).

It is possible to use OR to connect two statements. If either A OR B is true the combined statement is true. If an INCLUSIVE OR is implied, then either A or B or both may be true. Thus A OR B = 1 if A = 1 or B = 1 or both. Unfortunately, in order to obey the rules of algebra a Boolean sum is used, thus we write A OR B as A+B. The use of the plus sign has nothing to do with ordinary arithmetic addition; A+B is read as A or B. This can be confusing for a beginner. Other symbols have been proposed, but the plus sign has been universally adopted by computer engineers and we must therefore learn to live with it. The choice was not made lightly and there are good reasons for choosing the symbols used. When we make use of an EXCLUSIVE OR, i.e. when either one statement is true or the other statement is true but not both together, then we write the combined statement as $A \oplus B$. We often use the term NOT in connection with digital systems. For example, 'the safety cage is NOT closed' or 'the hopper is NOT full'. If we wish to negate a statement we put a bar across it, e.g. $\bar{A}$.

The main connectives AND and OR can be realized by using relays and associated switches. In Fig. 164(a) switches C and D are used. It may be that C = 'the safety cage is closed' and D = 'the hopper is full', and that a process can start only if C = 1 (i.e. it is true that the safety cage is closed) AND D = 1 (i.e. it is true that the hopper is full). If the process is dependent on the relay contacts Q being closed then, since Q = C.D, Q will be closed only if switches C AND D are closed.

In Fig. 164(b) we could have a light controlled by the switch Q on the relay. This light could be turned on if switch E is on OR if switch F is on; hence Q = E+F. This is an example of an inclusive OR since even if both switches are on, Q is on.

In a control system the type of switches used for the functions depends upon the nature of the function. C, the switch controlled by the safety cage, could be a microswitch and D some type of switch operated by a load cell. Figs. 164(a) and (b) also show what is called a **truth table** next to each of the circuit diagrams. This table shows all the possible combinations of C and D, and also

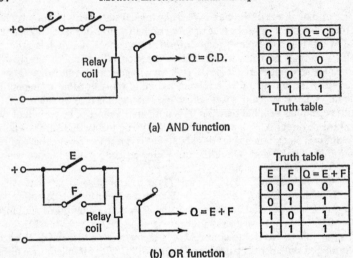

(a) AND function

(b) OR function

Fig. 164. Use of switches to perform the AND and OR functions. (a) Relay representation of the AND function. Relay contacts are shown in the unenergized position. (b) Relay representation of OR function.

E and F, together with the state of the outputs (Q) in the case of the AND and OR functions respectively.

Although not a connective (since it does not actually connect two Boolean statements), the function NOT is often used in digital systems. This function merely negates a system. Fig. 165 shows how it is realized in the case of

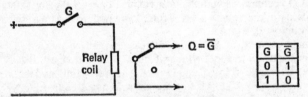

Fig. 165. Representation of the NOT function.

switches and a relay. If G is on then the relay contact Q is NOT on. The process of negation is called 'inversion' in computer jargon. There are numerous occasions when it is necessary to invert a function in control or computer systems. In Boolean terms, if $A = 1$ then $\bar{A}$ must equal zero since 1 and 0 are the only two truth values allowed. Hence if it is true to say that 'rain is wet' then it is not true to say that 'rain is NOT wet'.

The Transistor as a Switch

Switches that depend upon electromagnetic relays are suitable only for low-speed systems. Because of their physical size, low speed, vulnerability to wear in their mechanical moving parts and poor long-term reliability, relays are being replaced by semiconductor switches in digital systems. By comparison transistors are small, light in weight and inexpensive; they have no moving parts, an almost indefinitely long life when incorporated in well-designed circuits, and can switch at rates up to 1,000 million times per second. When they are incorporated into integrated circuits, many hundreds of transistor switches can be packed into an extremely small volume.

The ideal switch is one which has zero resistance between its terminals when in the 'on' position and an infinite resistance when 'off', and can switch from one state to another in zero time. There can thus be no power dissipated in a switch since either the current through the switch is zero when the switch is off, or there is zero voltage across the switch when it is on. We can operate a transistor in such a way as to approach the ideal switch performance sufficiently for our purpose. Fig. 166(a) shows how this can be done.

We have in effect a very basic amplifier, but can permit only one of two input signals. A is either at zero potential or it is at V_{cc}, the positive supply potential. Unlike an ordinary amplifier, no intermediate values are permitted but for the exceptions described below. Zero volts and $+V_{cc}$ volts can therefore be defined in terms of truth values. When $+V_{cc}$ corresponds to 'logic 1'

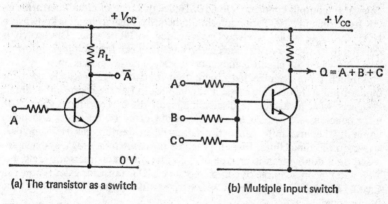

(a) The transistor as a switch

(b) Multiple input switch

Fig. 166. The use of the transistor as a switch. The multiple input switch acts as a NOR gate when positive logic is used.

and 0 V to 'logic 0' then we are said to be using positive logic. Since nearly all transistors in integrated circuits are *npn* transistors, positive logic—where the more positive voltage represents 'logic 1'—is the one used. However, there is nothing to stop us defining 0 V as 'logic 1' and $+V_{cc}$ as 'logic 0', in which case we would be using negative logic. In general, however, there is no point in adopting this second definition, since in the majority of cases everyone uses positive logic.

It will be noticed that when A is connected to the 0 V line the transistor is cut off; no current is passing and hence there can be no drop of voltage across the load resistor, R_L. The collector voltage is therefore at $+V_{cc}$. Alternatively, when A is at $+V_{cc}$ the circuit is so designed that sufficient base current is available to bring the transistor into saturation. The voltage across the transistor is then only about 150 mV. For logic purposes this is counted as being 'logic 0', even though the voltage is not quite zero. This is one of the exceptions mentioned above. The other exception concerns the collector voltage. When A is at 0 V the collector voltage is at $+V_{cc}$ provided the transistor is not connected to other transistor switches. The more transistors that are being driven the more base current is caused to flow through R_L and hence the collector voltage of the first transistor switch must fall. We must therefore relax our definition of 'logic 1'. Any voltage above an arbitrary percentage of V_{cc} is then defined as 'logic 1'. This may be 80 per cent, but varies somewhat depending on the type of electronic switch used.

In Fig. 166(a) it will be seen that when the input is at one logic level the output is at the other level. This is because of the 180° phase reversal in the transistor. In logic terms the circuit of Fig. 166(a) performs the NOT function. If two or more inputs are used, as in Fig. 166(b), then the transistor will be saturated if any of the inputs are connected to the $+V_{cc}$ line (logic 1). The output will then be at logic 0. Hence if A, B and C are all initially at 'logic 0' the output will be at 'logic 1'; if A OR B OR C is connected to 'logic 1' the output changes to logic 0. Hence this type of circuit performs an OR function and a NOT function simultaneously. Such a circuit is known as a NOT OR gate which is always abbreviated to **NOR gate**. The switch is referred to as a gate because such an arrangement can transmit logic information only when the circuit is in a specific state, i.e. the 'gate is open'.

A much more complicated circuit than that of Fig. 166(a) is used for modern integrated circuit gates. The complication arises because the circuit designers wish to avoid the disadvantages of the earlier simple gates. The main concerns are associated with speed of operation, reliability and noise immunity, i.e. immunity from acting falsely when spurious electrical impulses arrive on the input lines. For all practical purposes today logic systems are based on a double transistor type of gate (the transistor-transistor logic, i.e. **TTL gate**) or on a couple of complementary MOS transistor gates (known as **CMOS logic**). It is likely that CMOS gates will supplant TTL gates as the future standard gate.

Because we do not wish to draw out the complete electronic circuit every time we draw a circuit for a logic system, symbols for these gates have been adopted. Fig. 167 shows these symbols. Since so many of the gates that are used in the UK are manufactured by American companies, their symbols are also shown.

In all logic problems we must start with a specification outlining the requirements of the system. We may then draw a block diagram of the system and use this diagram to produce the most efficient system. Efficiency in this context means using the minimum number of logic gates to perform the function. Let us suppose, for example, that in a chemical factory the contents of a large vat must be pumped into a reaction chamber by an electrically driven pump.

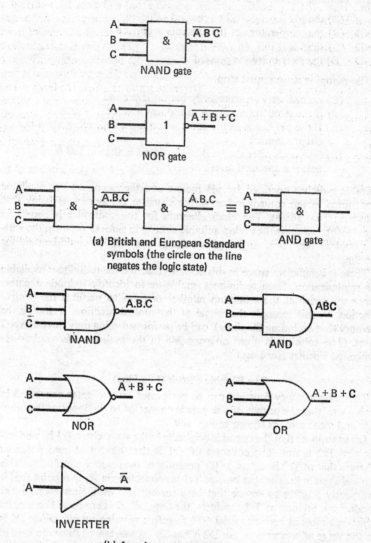

(a) British and European Standard symbols (the circle on the line negates the logic state)

(b) American gate symbols

Fig. 167. Gate symbols. The inverter using BSI symbols merely uses a single input NAND or NOR gate.

The pumping action must not commence unless

	(A) the pH has reached a specified value
AND	(B) the temperature of the contents is correct
AND	(C) the vat is full
AND	(D) the start button is pressed

The pumping action must stop

	(E) immediately an emergency button is pressed
OR	($\bar{B}$) if the temperature drops to too low a value
OR	($\bar{A}$) if the pH becomes too low (or too high, depending upon the circumstances)
OR	(F) 5 seconds after the level of the liquid in the vat has fallen below a specified level

Suitable switches operated by pH meters and thermometers are available from firms in the control industry. We visualize the system as being that shown in Fig. 168(a). The block diagram for the controller is shown in Fig. 168(b). We must then select suitable electronic gates to perform the various Boolean functions. An integrated circuit type NE 555 could be used for the timer.

There is insufficient space in this book to describe the techniques available for minimization. Such techniques enable us to identify redundant gates; hence we arrive at the minimum number of gates to perform the control function. In this example the whole of the control function, including the memory (but excluding the timer), can be performed using only three NAND gates. (The solution is given on page 348 in the author's *Electronics*—see Suggested Further Reading.)

The Bistable Circuit (Flip-Flop)

The basic memory function can be performed by two gates connected in such a way that the output of one gate is connected to the input of the other gate and vice versa, as shown in Fig. 169.

Let us suppose that the circuit is originally in the state where Tr1 is conducting and Tr2 is not. The collector of Tr1 is then almost at zero potential whereas that of Tr2 is almost at the potential of the positive supply line. This is a stable state because the base of Tr1 is connected via R_4 to a point that is sufficiently positive to ensure that base current is drawn, thus keeping Tr1 conducting. So long as Tr1 conducts, the base of Tr2 is connected to a point which is almost at zero potential. Tr2 therefore remains cut off. Diode D1 is on the verge of conduction, but D2 is heavily reverse-biased since the base of Tr2 is at about zero potential, and the collector of Tr2 is at a potential of almost $+V$. If now the trigger input is fed with square-wave pulses, a positive-going edge cannot have any effect on the state of the bistable because D1 is forced into the reverse-bias condition and D2 is already reverse-biased. The diodes D1 and D2 therefore effectively isolate the transistor bases from positive-going edges. A negative-going edge, however, brings D1 into conduction. Current is then diverted away from the base of Tr1 via D1 to C_3; this

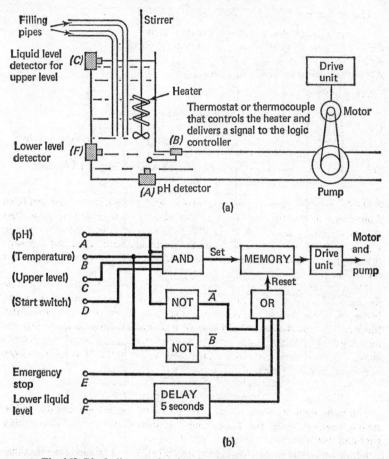

Fig. 168. Block diagram showing the solution to a control problem.

capacitor consequently acquires a charge. The reduction in the base current of Tr1 results in a fall of collector current. The collector voltage rises and Tr2 then starts to conduct. The accompanying fall in the collector voltage of Tr2 is transferred, via R_4 (and C_2, if present), to the base of Tr1. The base current is then rapidly reduced to zero and very quickly the bistable is driven into its other stable state in which Tr1 is cut off and Tr2 is conducting. Subsequently, the next positive-going edge discharges C_3 via R_7, R_3 and the low-resistance path provided by the base-emitter junction of Tr2, which is now in the conducting state. The discharge time is often reduced by shunting R_7 and R_8 with diodes, as shown in Fig. 169; the recovery time of the flip-flop is therefore improved. The bistable is now in the state whereby D2 is on the verge of conduction. The arrival of the next negative-going edge will initiate a rapid

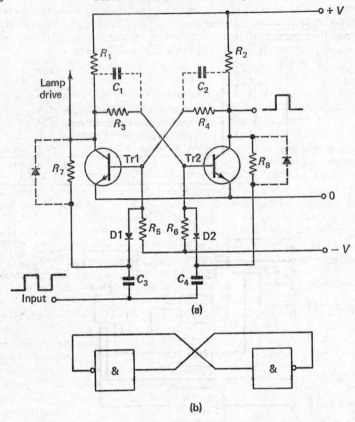

Fig. 169. The bistable circuit (flip-flop).

switching off of Tr2 and the switching on of Tr1. It will be seen then that the function of diodes D1 and D2 together with the a.c. coupling provided by C_3 and C_4 is to steer the incoming pulses in such a way that a toggle action is achieved. If negative pulses are applied to the input, then switching occurs on the leading edges of the pulse; the trailing (positive-going) edges discharge C_3 and C_4 alternately via the conducting transistor. If positive pulses are applied, switching occurs on the trailing (negative-going) edges; the leading edge of the next pulse then causes the discharging of the appropriate capacitor. Irrespective of whether positive or negative signals are applied to the input, it requires two input pulses to make the circuit go through a complete cycle. For every two input pulses there is thus one output pulse; the circuit therefore divides by two.

Improvements have been made on this circuit in order to be able to set and reset the voltages at the collectors. In computer work we also need to operate flip-flops synchronously, i.e. in step, and to be able to programme chains of flip-flops. These improvements result in a rather complicated circuit called a **JK flip-flop**. Such a circuit at one time could not be produced economically, but with the invention of the integrated circuit this is now a viable proposition. JK flip-flops are the standard bistable circuits used nowadays.

Counting

By connecting a chain of bistables in cascade we can achieve a divide-by-two action in each stage. In this way, if we wish to count the number of pulses in a train we feed them into the input of the chain of bistables and hence store the count. By having indicator lamps at the output of each stage the state of that stage is indicated. Thus the count is displayed as a row of lamps, some of which are on and others are not. If '1' indicates a lamp that is on and '0' represents a lamp that is off, then a count of eleven pulses would be indicated in the lamps as 1 0 11. Here the first bistable is on the right-hand side; the pulses are then fed into the right-hand side of the chain. In this way the least significant digit appears on the right and the most significant on the left. This is in line with our usual practice. The number 157, for example, in the usual decimal counting system uses three of the ten available digits (0 to 9). We adopt the convention of weighting the digits by multiplying them by 0, 10, 100, 1,000, etc., respectively. The least number is placed on the right, hence starting from the right we have the units, tens, hundreds etc. Hence 157 is actually 7 units plus 5 tens plus 1 hundred.

We adopt a similar system in electronic binary counting. Only two digits are available, viz. 0 and 1, and the weightings or multipliers are 2^0, 2^1, 2^2, 2^3, etc., i.e. 1,2,4,8, etc. When a number is stored in a chain of bistables it is not very easy for us to assess the magnitude of the number. This is because we have been brought up on the decimal system. Thus whilst we may have some idea of the size of a quantity expressed by the number 157 it is not easy to judge if 10011100 is greater or less than 157. For this reason we arrange that digital instruments such as clocks, voltmeters, frequency meters and computers give out information in the form of decimal numbers. The instruments themselves operate in the binary system because this is the most reliable electronic system. However, an interface is required that converts the binary number into decimal form.

BCD Numbers

It turns out that for digital instrumentation the pure binary system is not the most convenient to use. We therefore adopt a Binary Coded Decimal (BCD) arrangement. A chain of bistables is divided into groups of four. Each group of four represents a decimal number. Of course when we do this we cannot count as before by connecting one bistable output to the next input along the chain; special circuitry is necessary to count in the group of four bistables and then to transfer any 'carry' digit to the next group. Four bistables are necessary and sufficient to represent one of the ten decimal digits. One bistable has only two output states; two bistables can have a combination of four output states, viz. 00, 01, 10 and 11; three bistables can have eight states and this is just not sufficient for ten digits. We must therefore have four bistables.

Various weighting codes are used in connection with the four bistables in any group. Some are used for arithmetic processes, others for data transmission. The most common code used, however, is the 8421 BCD code. This is pure binary up to 1001 (=9), whereupon the four bistables must be reset to 0000 (=decimal 0) on the receipt of the next pulse. Most digital instruments nowadays give a BCD output (usually 8421 BCD output) in addition to displaying the information in decimal form. Such an output can be used in connection with code converters and other apparatus for data transmission and recording purposes.

Indicating Devices

After processing information, electronic equipment presents the results in a binary form to a **code converter**. The function of the converter is to generate signals that will energize indicating devices so that a number can be easily identified as being a series of decimal digits. Such binary-to-decimal converters can now be bought as large-scale-integrated circuits (LSI chips). At one time numerical indicator tubes (NIXIE tubes) were used, a glow discharge being established on an appropriate cathode shaped in the form of a decimal digit. Ten cathodes are mounted within the tube. Such tubes are still in use, but losing favour now since they require high operating voltages. For transistorized equipment voltage-compatible indicators are required. Three are in common use: the hot-filament display, the light-emitting diode (LED) type and the liquid crystal type. All depend upon the seven-segment arrangement shown in Fig. 170. The basic arrangement shown to the left of the diagram consists of seven segments in a figure 8 pattern. By energizing various combinations of segments different decimal digits can be represented. The example 539 is given.

Fig. 170. Seven-segment numeral display patterns used in connection with low-voltage numeral tubes that depend upon hot filaments, gallium arsenide phosphide diodes or liquid crystals. The basic arrangement is shown on the left; the pattern on the right shows how the number 539 appears.

Hot-filament displays employ a separate wire for each segment. When energized by the logic system, the wires glow in the same way as the filaments in an incandescent lamp. In spite of development, hot-wire displays consume considerable power and have a limited life. They are becoming obsolescent.

Lamps that can emit light continuously for many years are being developed by most of the big firms, notably RCA in America. These lamps are made from synthetically grown crystals of gallium phosphide and gallium arsenide phosphide. **Light-emitting diodes** (LEDs) made from these semiconductor compounds emit red or green light when a small electric current is passed through them. For all practical purposes almost no heat is generated. Low power consumption, long life, fast operating speed, small size and extreme reliability are the advantages that are gained when these light sources are used. For indicator purposes the diodes and diffusers are arranged in the seven-segment pattern.

Liquid crystal displays form the third group of indicator devices. Liquid crystals are organic substances that are termed 'mesamorphic', being intermediate between the solid and liquid state. Various types of liquid crystals are known, but the type of use in indicators can be controlled electronically. In the undisturbed state, thin layers of this material appear to be quite transparent. The backing for the liquid is black so that incident light is absorbed. For display purposes the thin (about 0·002 inch) layer of liquid is sandwiched between a sheet of conductive glass and a layer of plastic. The plastic film has the numeric pattern cut away, as shown in Fig. 171. Underneath the pattern are areas of gold or other conducting material arranged in an identical

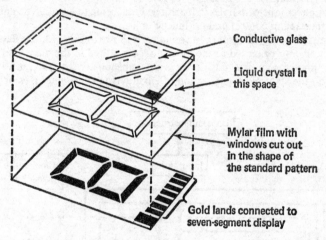

Conductive glass

Liquid crystal in this space

Mylar film with windows cut out in the shape of the standard pattern

Gold lands connected to seven-segment display

Fig. 171. Exploded view of a liquid crystal numeral tube. The actual thickness of the liquid crystal film is about 0·002 in. The whole sandwich is hermetically sealed. Gold end contacts are exposed and connect the pattern and conductive glass to the logic system.

Fig. 172. A typical clock display. The initial 'dot' shows whether a.m. or p.m. is being displayed. The colon pulsates at one-second intervals; if only one part is illuminated the alarm is off, but if both 'dots' of the colon are displayed the alarm is set.

pattern. Only the wanted areas are then activated. This activation is achieved by applying a small voltage across the area of liquid of interest. The molecules then rearrange themselves so as to form a milky liquid. Incident light is reflected in much the same way as it is from milk, and the energized portion of the liquid appears to be white and diffuse.

Digital Computers

We can do no more here than touch upon this very wide subject. The computer is a large organization of logic systems that can perform numerical calculations very rapidly. After making the calculations the results can be displayed on cathode-ray tubes in pictorial form, or through a typewriter to yield the results as a permanent record. Alternatively, the output mechanisms can be made to operate machines. The computer uses a special language which is binary in nature. It is necessary to feed in the information from punched cards or paper tape with holes in it arranged according to a code. The data are then processed in control and arithmetic units.

A simplified block diagram of a computer is shown in Fig. 173. The data and instructions, on being fed to the input unit, are stored in the main computer store. Although any two-state device is a possible binary store, so far we have relied on the state of magnetization of small ferrite rings and tapes or discs

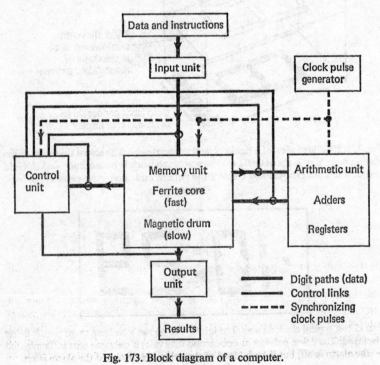

Fig. 173. Block diagram of a computer.

coated with iron oxide. In the latter case, information is stored on the disc using similar techniques to those used in tape recording. The main core store is built from thousands of ferrite rings; not only can they be magnetized by wires threaded through the rings, but the state of magnetization can be detected. Magnetizing the rings is called 'writing' and the detection of the state of magnetization is termed 'reading'. Information is often stored on a back-up store or disc. Writing and reading is undertaken by tape recorder type heads. Fig. 174 shows these types of store.

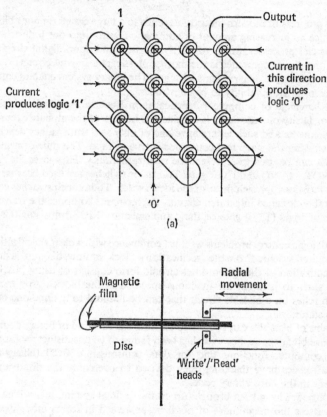

Fig. 174. Two types of magnetic data store: (a) ferrite ring memory store; (b) back-up disc magnetic store.

When an arithmetic operation is to be performed the control unit reads the data in the store, i.e. addresses the store, and transfers the relevant portions of it to the arithmetic unit. This unit contains registers which store the data temporarily. The control unit then transfers the data to shift registers and the clock pulses move the data digits along the shift registers into the adders to be added. The result is then moved into another register and back into the store. The whole sequence of operations must be synchronized, which is the reason for having a square-wave oscillator called a clock pulse generator. The gates in the system are inhibited until the clock pulses are present, whereupon the digital data can be processed all at the correct time. The shift registers consist of chains of flip-flops; nowadays all the necessary flip-flops for a register are fabricated on a single chip.

Summary

The progress of modern industry depends to a large extent on our ability to introduce an increasing amount of automatic control into our industrial and commercial processes. Such control systems depend upon **digital circuits**, of which the digital computer is the most sophisticated. Digital circuits depend upon two-state (i.e. **binary**) logic switches. The binary system enables complex systems of great reliability to be constructed.

The language of computers depends upon **Boolean algebra** in which the truth or falsity of statements is considered. Letters of the alphabet represent the statements and such letters have one of only two truth values depending upon whether the statement concerned is true or false. The state of **transistor switches** can be used to represent the two truth values. Thus arise the AND, OR, NOT, NAND and NOR gate. Transistor switches are used because their characteristics approach those of the ideal switch. Today such switches consist of suitably designed integrated circuits. Such circuits comprise the transistor-transistor logic (TTL) gate or the complementary MOS transistor (CMOS) gate.

In all logic control problems we must commence with a clear specification of the required system. From the specification a block, or flow, diagram is drawn. The realization of the system in electronic form consists of using NAND or NOR gates to perform the functions specified in the blocks. Any memory system relies on a **bistable circuit** that can be induced to assume one of two stable states.

Chains of bistables can be used to perform the function of binary counting. The pure binary code, however, has been found to be unsatisfactory for arithmetic computer functions and for data transmission. BCD (binary-coded decimal) codes have therefore been devised to overcome the disadvantages inherent in the pure binary code.

Because we have been brought up on the decimal system, it is not easy for us to assess the magnitudes of numbers expressed in forms other than the decimal characters. An interface problem is therefore created between man and machine. The digital machine uses the binary system so as to achieve a high degree of reliability. However, the binary data must be converted into decimal form so as to be readily interpreted by human beings. Various electronic visual display devices have been invented to enable a rapid appreciation of the data.

Digital computers are the most sophisticated of any form of logic circuitry. Data and instructions, previously stored on punched cards or tapes, are fed into the computer. This information instructs the computer to perform specified operations on the data supplied. For various technical reasons different 'languages' have been devised to make easier the task of human beings when 'talking' to the computer. The computer stores its information as binary digits (or bits) in bistables. The bistables of computer memories are usually ferrite rings or magnetic disc stores.

Questions

1. Why do we need the special language of Boolean algebra when dealing with logic circuits?

2. Why are transistors the only suitable devices available for large-scale computer and logic circuitry?

3. Explain the action of a bistable circuit.

4. Why are digital computers necessary to a modern technological society?

5. Why is it desirable that computers are of the digital type?

6. Is the binary number 10011100 greater or less than the decimal number 157?

Suggested Further Reading

Jacobowitz, H., *Electronic Computers Made Simple*, Heinemann, 1976.

Olsen, G. H., *Electronics: A Handbook for Engineers and Scientists*, Butterworth, 1973.

Olsen, G. H., *Computers and Microprocessors Made Simple*, Heinemann, 1980.

Reddihough, J. A., and Sinclair, I. R., *Beginners Guide to Transistors*, Newnes-Butterworth, 1975.

14

PHOTOELECTRIC DEVICES

There are many occasions on which it is convenient to replace the human eye and operator with some sort of automatic device. In gas- or oil-fired boilers and furnaces, for example, where flame failure would constitute a serious hazard, it is not economic to employ a man solely for the purpose of detecting flame failure and raising the necessary alarm. Such detection and subsequent action can be carried out more cheaply and reliably by electronic means. In applications such as burglar alarms, fire alarms, signalling and warning systems, smoke detection, counting and automatic control, colour comparison, etc., the human operator can now be replaced by reliable electronic apparatus. Such apparatus depends upon photoelectric devices. The devices, often called photocells, have the ability to convert light into corresponding electric signals. These electric signals can then be processed by conventional electronic equipment to produce the desired results.

Photoelectric cells can do more than merely replace the human eye. Because of their ability to measure light intensity accurately, and to respond to radiations in the infra-red and ultra-violet regions of the electromagnetic spectrum, photocells are superior to the human eye. Using them as the detecting agency it is possible to build apparatus for specific purposes; infra-red spectroscopy, the estimation of turbidity, fluorimetry studies and photometry are some of the examples in which photocells can be used for scientific work.

Photoelectric devices may be classified according to the mechanism by which they operate. The three main classes of photocell are: (a) photoemissive cells, (b) photoconductive cells and (c) photovoltaic cells.

Photoemissive Cells

It has been discovered that metals such as caesium, sodium, potassium and rubidium emit electrons when subjected to electromagnetic radiations of short wavelength such as ultra-violet radiations or visible light. The electrons are more readily emitted when the surface is contaminated with films of foreign elements; these films must have a thickness of only one or two atoms or molecules—hence they are extremely thin. In order to collect the electrons as useful current, the emitting surface, called the **cathode**, must be enclosed in an evacuated glass tube. The electrons are then able to travel towards another electrode, called the **anode** because it is held at a positive potential. At one time these cells were the standard light-measuring cells, and were also used to produce signals from the sound track of entertainment films. To a very large extent they have been superseded by solid-state devices. One type of photoemissive device, however, has not yet been replaced; this device is the **photomultiplier tube**.

278

Photomultiplier tubes are photoemissive cells of great sensitivity. Strictly they are misnamed, since multiplication of light does not occur. This type of tube should be called an electron-multiplier tube. However, the term 'photomultiplier' is in common use.

Electron-multipliers have photoemissive cathodes which produce electrons when light falls upon them. These electrons are then accelerated by a system of anodes, called **dynodes**. Each dynode along the system is maintained at a progressively higher voltage; the usual voltage between adjacent dynodes is about 100 V. Each dynode is coated so that when a primary electron strikes the electrode, secondary electrons are emitted. Between two and three electrons on average are produced by the impact of a single electron. Since the tube may have between 10 and 13 stages it will be realized that considerable electron-multiplication occurs. The sensitivity of the tube is therefore enormous.

To prevent the electrons emitted from the photocathode and dynodes from travelling directly to the final anode, the dynodes are specially shaped. Figs. 175 and 176 show two common arrangements. In the circular type of electron-

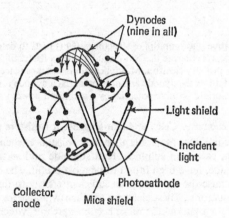

Fig. 175. Structure of a circular photomultiplier tube.

multiplier the dynodes are specially shaped so as to produce electric fields which direct the electrons to each dynode in turn. In the 'venetian blind' arrangement each dynode consists of a layer of electrodes inclined so as to produce electron paths as shown in Fig. 176.

The advantages of the electron-multiplier lie in the very high sensitivities which are obtained without recourse to an associated high-gain amplifier. The signal-to-noise ratio is therefore much better than it would be if conventional amplifiers were used. The linearity, stability and frequency response are very good. Typical applications are the counting of particles in nuclear physics studies (where the particle gives rise to a scintillation, i.e. flash of light, in a suitable crystal, e.g. sodium iodide) and, as we have seen, in television camera tubes.

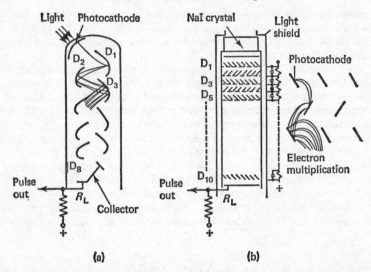

Fig. 176. Alternative photomultiplier structures. (b) is used to detect nuclear radiations. The particle, on entering the NaI crystal, causes a flash of light which initiates the release of the primary electron. Multiplication then takes place as the secondary electrons proceed down the dynode arrangements. Each dynode is maintained at the appropriate positive potential by a stabilized power supply.

Photoconductive Cells, or Light-Dependent Resistors (LDRs)

Certain semiconductor materials alter their resistance when light falls upon them. Selenium, cadmium sulphide, lead sulphide and selenide, as well as indium antimonide, have been found to be most useful. The change in resistance occurs because the light supplies sufficient energy to dislodge electrons from their parent atoms. These electrons are then available for conduction; the conductivity of the material therefore becomes greater, which is just another way of saying that the resistance is reduced.

The usual construction of a photoconductive cell is shown in Fig. 177. The semiconductor material is deposited as a layer a few microns thick on to the

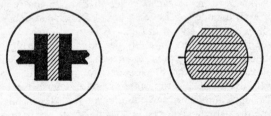

Fig. 177. End-on views of the electrode structure of two forms of photoconductive cell. The semiconductor material is shown shaded between the metallic contacts. The structure on the left is used for spectroscopic work.

electrodes (1 micron = one millionth of a metre; 25 microns are approximately equal to 0·001 inch). The arrangement on the left of Fig. 177 is intended for spectroscopic work whilst that on the right is for general work. The substrates are usually of glass, and the whole cell is sealed into a protective mount.

Applications of Photoconductive Cells

Since the resistance of this type of cell changes in accordance with the strength of the incident light, obvious applications involve using the cell to alter the bias current into a transistor. In this way the transistor collector current then depends upon the illumination of the cell. Fig. 178, for example,

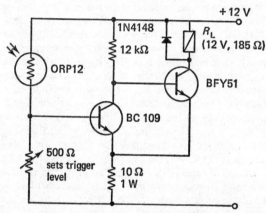

Fig. 178. Light-controlled Schmitt trigger suitable for use as a flame-failure device, automatic illuminator or for counting objects that break a beam of light which illuminates the photocell. Relay type is not critical; it should operate at voltages between 8 and 12 V. RS Components type 40 is satisfactory. The diode is required to protect the transistors. When the relay is de-energized a large back e.m.f. is produced which could damage the BFY51; the diode prevents the generation of a large back e.m.f. Diode should be type OA47, 1B30 (RS Comps.) or 1N4148.

can be used for flame failure, as an automatic parking light, or as a counter. The Schmitt trigger is a special circuit in which the final load current is suddenly turned on or suddenly turned off. In our example, when the collector voltage of the BC109 falls (because the cell is illuminated, thus increasing the base current to the BC109), the current in the BFY51 falls. Hence, since there is now less current flowing in the 10 ohm resistor, the emitter voltage of the BC109 falls. This turns the BC109 on harder. The increase in the collector current of the BC109, and hence the 10 ohm resistor, does not, however, make up for the reduction of the current in the BFY51. There is thus a regenerative reaction whereby the BFY51 is suddenly turned off. When the light is removed a comparable snap action turns the BFY51 on. R_L can be a relay with a changeover contact; a wide variety of alarm and control functions can therefore be performed with this circuit.

Photovoltaic Cells

In this type of cell, illumination falling on the device generates an e.m.f. between the cell's terminals. The phenomenon is known as the **photovoltaic effect**. The cell consists of a metal–semiconductor or semiconductor–semiconductor arrangement; it is the potential barrier which is formed at the respective junctions that provides the mechanism for the generation of the e.m.f. within the cell.

Readers will recall from Chapter 3 that when a *pn* junction is formed, a migration of electrons and holes takes place. The electrons are trapped in the acceptor impurities near the junctions. The electrons have come mainly from the donor impurities on the other side of the junction, but we find it convenient to say that holes have been trapped by the pentavalent impurities in the *n*-type material. A barrier layer is thus established. If we reverse-bias the *pn* junction no load current flows, but careful measurement reveals a small, but detectable reverse leakage current. This current is due to the formation of electron-hole pairs within the barrier layer. These electron-hole pairs result from the breaking of a valence bond between a couple of adjacent semiconductor atoms. The energy required to disrupt this bond comes from thermal sources, mainly the heat stored in the air around the component or the wire leads connected to it. If we manufacture the *pn* junction in such a way that light or other radiation can penetrate the barrier layer, then once again electron-hole pairs are formed; the energy in this case comes from the light, or radiation, source. This energy can be used to move the electrons away from the *pn* junction and around an external circuit; thus even without any externally applied voltage we can generate an electrical current, and so convert light energy directly into electrical energy.

To be effective the junction areas of the photocell must be quite large compared with those of an ordinary rectifier. The voltage generated within the cell depends upon the materials used to manufacture the *pn* junction, the intensity of the illumination and to some extent on the current flowing in the external

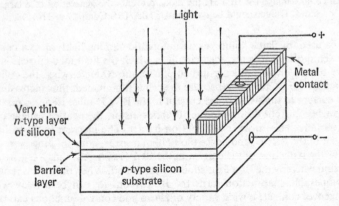

Fig. 179. Silicon photovoltaic cell (solar cell).

circuit. For **silicon** cells the open-circuit voltage may be as much as 0·5 volt. Silicon photocells are manufactured by diffusing a highly doped layer into a substrate that is lightly doped. The substrate therefore has a higher resistivity than the diffused layer. Such a diffused layer must be thin enough to be almost transparent in order that light can penetrate easily to the barrier layer formed at the junction. Fig. 179 shows a **solar cell** as a source of power.

Materials other than silicon are also used. Before the development of silicon technology the semiconductor **selenium** was in common use. Such cells are manufactured by first roughening a steel or aluminium base plate and coating the surface with bismuth or some other metal known to form a non-rectifying (i.e. ohmic) contact. Selenium is then deposited under vacuum and later annealed so the original semi-amorphous material is converted into its crystalline semiconductor form. An extremely thin, transparent inert metal such as gold is deposited on the selenium surface. A metal ring then forms the top contact, the bottom plate being the other connection. The spectral response of the selenium cell is similar to that of the human eye; this type of cell is therefore still popular in colour comparators and for certain photographic work.

Other semiconductor materials in common use today are **cadmium sulphide** and **cadmium selenide**. So far the efficiency of these cells has not been high. A silicon cell, for example, has an efficiency of only about 8 or 9 per cent—that is to say, only 8 or 9 per cent of the radiant energy falling on the cell is converted into electrical energy. However, thin-film cells based on **gallium arsenide** are very promising. Sample cells produced by the American Space Agency (NASA) have shown efficiencies of 15 per cent. The Jet Propulsion Laboratory team predict that cells are likely to be developed with efficiencies approaching 19 per cent. Space research, of course, depends heavily on photovoltaic cells since by arranging many of them on panels attached to a satellite they form the power supplies for the electronic equipment.

Although a silicon photocell is basically a photovoltaic cell, in the absence of light it behaves as a normal *pn* junction. If a reverse bias is applied, the normal reverse characteristic is involved. In darkness the cell exhibits a very high resistance. If, however, under reverse-bias the cell is illuminated, the resistance falls; thus it is possible to use such a cell in a photoconductive mode. Small diodes of this type are often used as read out devices in computer systems.

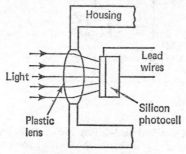

Fig. 180. Silicon photodiode.

As the punched tape or card passes over an array of cells some of them will be illuminated when holes in the tape or card coincide with the position of the appropriate cell. Fig. 180 shows the construction of such a cell, which may be a point-contact type or a *pn* junction type. Often plastic lenses are incorporated into the housing so as to concentrate the light on to the active region of the cell.

The Phototransistor

Phototransistors are devices that combine the photovoltaic effect with transistor amplifying action. They are housed in metal containers of the type used for conventional transistors except that a hole is cut in the metal to admit light. Illumination of the collector-base depletion layer creates electron-hole pairs. For an *npn* device the electrons are swept into the collector region and the holes into the base. These holes constitute a 'signal' current which is amplified as described in Chapter 3. In a sense the transistor cannot discriminate between holes entering the base lead from an electrical signal source or holes entering the base region from some other source such as the collector-base depletion layer. The sensitivity can be improved by having a Darlington pair arrangement in a single package, a typical example of which is the Ferranti ZM100. Some typical circuits using phototransistors are shown in Fig. 181.

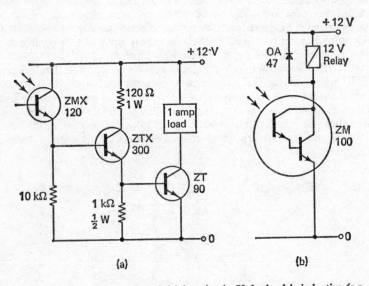

Fig. 181. (a) shows a standard load-driving circuit. If the load is inductive (e.g. a relay) a diode must be used as in (b). (b) Darlington phototransistor driving a relay directly. In both cases no connections are made to the base.

Summary

Photocells are transducers that convert light into corresponding electrical signals. They can replace the human eye in many instances when alarm or control functions are involved, e.g. burglar alarms, fire alarms, smoke detectors, colour comparators, automatic control systems, etc.

Photocells can be classified as photoemissive, photoconductive or photovoltaic. In the **photoemissive** types light falling on the cell releases electrons into a vacuum. These electrons are subsequently collected by an anode.

In **photomultipliers** several special anodes are used so that secondary electrons are produced when primary electrons bombard the anode surface.

In **photoconductive** cells the resistance of the cell depends upon the intensity of the light falling onto the active surface. The change of resistance can be made to alter the operating conditions in an electronic circuit so as to effect an alarm or a control function.

Photovoltaic cells generate voltages when subjected to radiations in the form of light. Cells made from silicon or other semiconductor material can convert sunlight directly into electrical energy; they are then known as **solar cells**.

The **phototransistor** is a device that combines the photovoltaic effect with transistor amplifying action.

Questions

1. How important are photocells in modern technology?

2. How does a phototransistor work?

3. Would you select a photoconductive cell or a phototransistor for burglar alarm purposes?

4. The electron-multiplier (photomultiplier) seems to be the only type of photoemissive cell in common use today, although other types were at one time very popular. Can you account for this?

15

PROJECTS

There is no doubt that practical work involving the building of electronic apparatus does much to further one's understanding and enjoyment of the subject. No amount of reading alone will make a person competent to deal with electronic equipment; practical experience is the only way to do this. The reader should therefore take every opportunity to do practical work. One excellent way of starting is to buy a kit and build a transistor radio or an audio amplifier. The end products will have more than just an educational value. Several firms now market kits of all kinds. The name of a firm that supplies kits and components is given at the end of this chapter. It is essential to develop practical skill, experience and perseverance. On no account must the reader become discouraged if the apparatus he builds does not work the first time it is switched on. Finding the faults in apparatus can lead to an increased knowledge of the subject.

The amount of work undertaken will depend upon the available time, the amount of money available and the availability of a suitable place to work. It is surprising just how much can be accomplished by the keen amateur on a very limited budget, using the kitchen table as a workbench. Buying kits then becomes a very economical way to purchase components since there are no wasted parts.

However, for those lucky enough to have their own place to work and store gear it may be worth while setting up a little laboratory. A stock of resistors and capacitors can be obtained very cheaply. So far as semiconductors are concerned a great deal of work can be done with a very limited range. In many circuits almost any type of silicon transistor works satisfactorily. Even when a type is specified on a circuit diagram a general-purpose device will do. The initial range can be limited to OA91 (a point contact diode), 1N4001 (a rectifying diode for voltages up to about 50), 2N3702 (*pnp*) and 2N3704 (*npn*) transistors and a 741 integrated-circuit operational amplifier. So far as power transistors are concerned, it is better to use the specified type until experience has been gained.

A basic set of tools will also be required. This should comprise a soldering iron, a pair of wire strippers, a wire cutter, small screwdrivers and a pair of thin long-nosed pliers.

Many of the circuits that beginners build can be assembled on the printed circuit boards supplied with the kits. Where these are not available Veroboard is a good alternative. This is an insulating board to which copper strips have been bonded. A standard matrix or pattern of equally spaced holes is drilled along the copper strips. The hole-spacings available are 0·2, 0·15 and 0·1 inch. The 0·1 inch spacing is to be preferred as a standard, since this is the standard spacing of the pins on a dual-in-line integrated circuit. The components are assembled on to the board on the opposite side to the copper strips. These components are held in place by pushing their connecting wires through the holes and soldering them to the copper strips. Provided some skill is developed in designing the layout, the copper strips form the majority of the interconnections between the components. The remaining connections are made by using links of insulated wire. The strips can be broken at convenient positions by using a small drill.

Soldering is an activity that requires some care if poor results are to be avoided. Poor electrical soldering is the cause of many troubles in beginners' apparatus. The correct iron and solder for the job are essential. For most wiring work a 25 W iron will be found to be satisfactory, although with components such as transistors, a 15 W iron with a pencil bit will prevent excessive damage. It is a matter of experience to judge when soldering conditions are right. On the one hand, the majority of unsatisfactory joints are caused by dirt and too little heat, but on the other hand excessive heat can burn the insulation on wires, bring the copper away from printed circuit boards or damage semiconductor components. The application of a hot iron for a short period should be the general rule rather than a 'cool' iron for a long time. The bit of the iron should always be kept clean and tinned. If the bit is pure copper it will be found that the tip is eaten away during use; in fact, the copper is dissolved by the solder. It will be necessary to file the tip flat at an angle of about 45° from time to time, and then re-tin. Nowadays soldering-iron bits that have been specially passivated with an iron coating have become available. Corrosion is therefore eliminated. Such tips must never be filed.

Solder with a high tin content is necessary for good results. Ersin multicore solder is excellent for electronic work. The tin content is 60 per cent (the remaining 40 per cent being lead), and this solder is sold as reels of 'wire' in which the flux is incorporated as a core. For the joint to be firm and sound the solder must 'wet' the surface. A good mechanical joint must first be made by wrapping the wire around or through a tag or, in the case of two wires, these should be bent over so one wire is hooked on to the other and then squeezed. The iron should then be applied to the joint and then the solder wire applied. The solder should then be allowed to flow evenly over the copper strip and wire, tag and wire or two wires, depending upon the type of joint.

It is almost impossible to recommend items of test gear since so much depends upon the available financial resources. Some sort of voltmeter is essential, however, with a milliammeter coming a close second. An AVO 8 is an excellent multipurpose instrument, but unfortunately these multimeters are very expensive (about £60–70 in 1977). Two of the projects given later in the chapter are worth considering in this connection. Not only do the items made in the project provide the necessary measuring equipment, but building the instruments can provide useful experience and satisfaction as well as saving money.

Some sort of oscillator is useful; this again can be built for oneself using the circuit given later. The usefulness of an oscillator is increased enormously if an oscilloscope is at hand, but not many amateurs can afford to buy their own oscilloscope in the initial stages. However, it should be stressed that almost all the measuring that is required by a beginner can be performed by a multi-meter (either home-made or bought).

Circuit diagrams are given below of items of electronic apparatus that should prove useful and interesting. The lead configurations of the transistors and integrated circuits used are given at the end of the chapter.

Project 1: A Simple Radio Receiver

Fig. 182 shows the circuit diagram of the simplest possible radio receiver. The radio waves are concentrated within a ferrite rod around which is wound

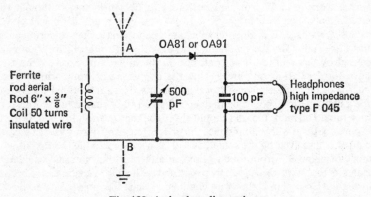

Fig. 182. A simple radio receiver.

50 turns of thin insulated wire. The type of wire is not important, but the turns should be close together and all wound in the same direction. The inductance so formed makes a resonant circuit with the variable capacitor. By adjusting the variable capacitor, the resonant frequency can be varied; thus we can tune in various medium-wave stations. Demodulation is achieved by the diode and fixed capacitor; the headphones then convert the audio electrical signals into an audible sound. The receiver should be used within an area where the radio signals are strong. A length of aerial wire attached to the point A and an earth connection (the central heating or cold water pipe—but *NOT* a gas pipe) at point B will increase the signal strength.

Project 2: A Variable-Gain Amplifier

Using an integrated circuit to provide us with amplification saves us a lot of time because the various transistors and resistors have already been assembled by the manufacturer. Operational amplifiers must always be used with feed-back otherwise the very high open-loop gain would give us some problems in keeping the amplifier stable. The 741 type of integrated circuit amplifier is easy to use because it is internally compensated, and thus is always stable in operation when used as a feedback amplifier.

Readers will recall that the gain of the arrangement shown in Fig. 183(a) is $-(R_f/R_i)$. If instead of feeding R_f from the output terminal we feed it from a tapping on a potential divider chain, as in Fig. 183(b), we reduce the feedback and hence increase the gain. The circuit shown allows us to vary the gain from 1 to just over 200.

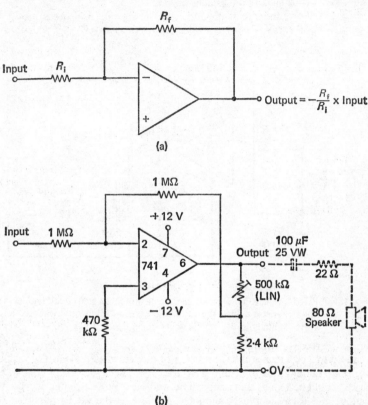

Fig. 183. A variable gain amplifier with a gain variation of 1 to just over 200. The input impedance is 1 MΩ. The additional components (shown dotted) can be used to allow audio signals to be heard. The input terminals could, for example, be connected in place of the headphones in the circuit of Fig. 182. The integrated circuit is a 741 from any manufacturer. Type numbers may be μA741 or SN72741, for example.

Project 3: An Integrated-Circuit Radio

It will be found that the simple receiver of Project 1 is not very sensitive or selective. A great improvement can be achieved by using R.F. amplification. Tuned R.F. (T.R.F.) receivers of high sensitivity can be made from discrete components, i.e. resistors, capacitors and transistors. Fortunately for the amateur this is not now necessary since a complete A.M. tuner can be obtained in integrated circuit form. The Ferranti ZN414 is a good example. In order to build the tuner the only additional components required are a ferrite rod, three fixed and one variable capacitor and a couple of resistors. If high-quality, low-impedance headphones are used, an inexpensive load-matching transformer will also be required. Alternatively, the power amplifier of Project 4 could be used to give a loudspeaker output.

Fig. 184 shows the circuit diagram. The ferrite rod, 6 inches in length, is wound with 50 turns of wire as in Project 1. Alternatively, if it is desired to

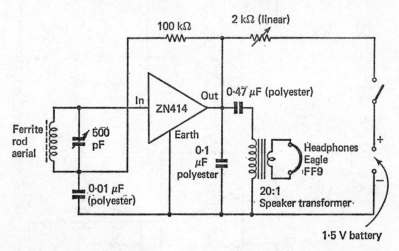

Fig. 184. A T.R.F. radio receiver based on the Ferranti integrated circuit type ZN414. The 2 kΩ variable resistor adjusts the automatic gain control facility; it should be set initially in the middle of its travel, i.e. at 1 kΩ.

make the receiver within as small a volume as possible the rod length can be reduced to 2 or 3 inches and the number of turns increased to 60. The variable capacitor should then be a solid dielectric type. The layout is not critical, but leads should be kept as short as possible, and the 0·1 μF capacitor between the output and earth should be mounted as close to the i.c. as possible. The transformer may be almost any small speaker transformer. The ratio of primary turns to secondary turns is not critical and may be about 20:1.

Project 4: A Power Amplifier

Where more power is required to operate a loudspeaker than is available from the i.c.s of Projects 2 and 3 we must use a power amplifier. Many circuits have been published for high-power stereo amplifiers, but the beginner should gain experience before attempting their construction. A modest alternative that gives a good output volume is an amplifier based on the circuit of Fig. 185. Although not a hi-fi amplifier in the true sense of the term, this circuit will perform very satisfactorily. The output is a push–pull arrangement based on a complementary pair. This pair of output transistors should be bought as a matched pair from a reputable dealer.

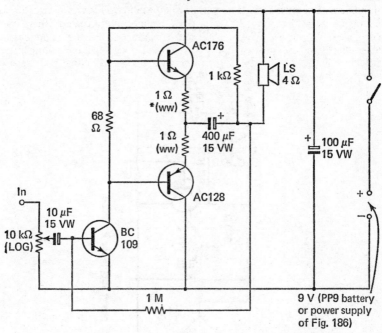

Fig. 185. A transformerless push–pull audio power amplifier.

Project 5: A Mains Power Supply

So many different supply voltages are used for electronic apparatus that it is not possible to design a universal power supply. However, the basic circuits can be modified easily enough to give the desired results.

In Fig. 186 we have a stabilized mains supply suitable for the power amplifier of Project 4. The output voltage is controlled by the zener diode. In our case, since we require a supply voltage of about 9 V for the power amplifier, a 10 V zener is chosen. Bearing in mind the fact that about 0·6 V is dropped across the base-emitter junction of the 2N3704, the output voltage is then about 9·4 V. We can easily see how to obtain different output voltages. For any given output voltage a zener voltage is chosen about 0·5 V above the required output voltage. For example, if we required an output of 12 V then a 12·5 V zener would be needed. Since this is not available a 13 V zener diode would be used. The output would then be about 12·5 V, but this would be quite satisfactory in the majority of cases. No modification to the circuit is necessary provided the output voltage required does not exceed about 15 V. For greater voltages the secondary of the transformer would have to deliver more voltage. The r.m.s. voltage of the secondary should be about 90 per cent of the maximum d.c. voltage required. The 2N3704 will deliver up to about 800 mA. For greater currents a power transistor will be needed. A 2N3055 can be obtained cheaply nowadays and this will pass currents of several amperes with safety.

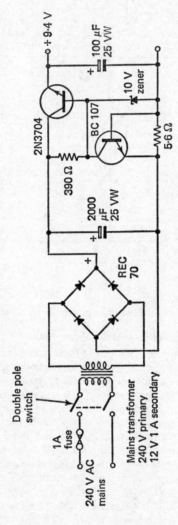

Fig. 186. Stabilized power supply for Project 4.

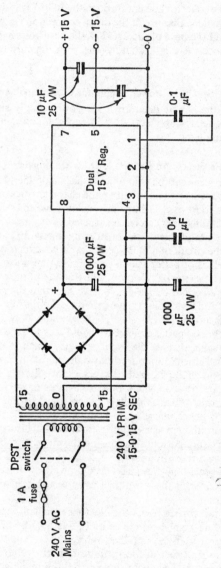

Fig. 187. A stabilized power supply for any project that uses a 741 i.c. The dual 15 V regulator is an RS Components type 305–636.

An interesting feature of this circuit is that we have short-circuit protection. If a short on the output terminals does occur the current through the 5·6Ω resistor increases the collector current through the BC107 transistor and hence the collector voltage falls. Because of the direct connection between the collector of the BC107 and the base of the 2N3704, the base voltage of the 2N3704 is reduced and hence there is a limiting effect on the current through this transistor.

Project 6: A Stabilized Power Supply for Operational Amplifiers

Fig. 187 shows the circuit diagram for a power supply suitable for the majority of operational amplifiers. It is intended for use in projects that employ the 741 amplifier.

Project 7: An Electronic D.C. Multimeter

We have already seen that the gain of a feedback amplifier depends only on the feedback components and not at all on the characteristics of the amplifier, provided the open-loop gain is large. By using close-tolerance high-grade resistors a very stable amplifier can be constructed having an accurately known gain. We can make use of this when constructing measuring instruments.

Fig. 188 shows the circuit of a multimeter that is not expensive to build and yet performs as well as or better than a commercial multimeter costing much more. Fig. 188(a) shows the basic circuit. The fraction of the output fed back to the input is $R_2/(R_1+R_2)$; this is β. We have seen that the gain of such an arrangement is $1/\beta$ and hence the voltage gain if we put the input into the non-inverting (+) input terminal is $(R_1+R_2)/R_2$. By selecting only a portion of the input voltage, using a potential divider, we could measure the output voltage and, knowing the gain, deduce the input voltage (which is the one we are measuring). For instrumentation purposes we do not use a carbon track type potential divider, but prefer instead to select points along a chain of high-grade resistors by means of a switch. For fixed values of the resistors R_1 and R_2, the current through them is directly proportional to the voltage across them (Ohm's law); hence we can use a moving-coil milliammeter as our indicator to measure the current through R_1 and R_2. This arrangement can also be used to measure current. By causing the current to be measured to pass through a known resistor, a known voltage is developed. The meter can then be calibrated accordingly. A large-scale drawing on graph paper can be attached to the box housing the instrument; a suggested form of the drawing is shown in the diagram.

We have not so far worried about the fact that even with zero input voltage the output voltage may not be quite zero. This is because of inevitable imbalance in the stages within the amplifier. Whilst this is of no consequence for audio and some other types of work, indication errors are introduced in measuring equipment. To eliminate these errors the manufacturer has provided facilities for compensating for the imbalances; these lead to the zero-set control connected to pins 1 and 5.

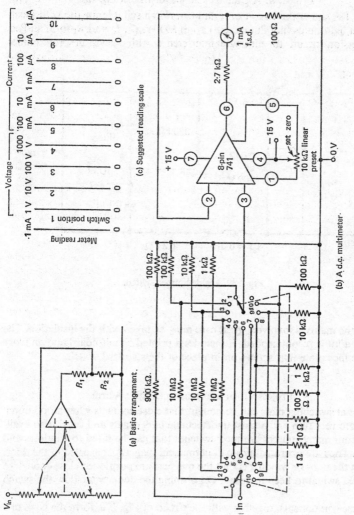

(a) Basic arrangement.

(b) A d.c. multimeter.

(c) Suggested reading scale

Fig. 188. An inexpensive d.c. multimeter.

Project 8: A Dimmer Circuit for Mains Lighting

Fig. 189 shows the circuit of a dimmer switch suitable for use with incandescent light bulbs (not fluorescent) up to 300 W. C_1, L_1 and R_1 form an R.F. suppression circuit to eliminate interference with television programmes.

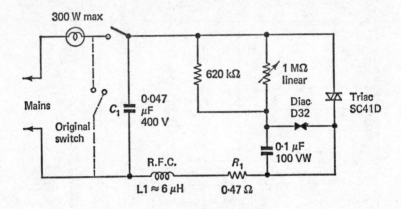

Fig. 189. Light dimmer switch.

Since the mains is involved, great care must be taken with the insulation. The use of a kit is recommended; it includes a printed circuit board and an ivory plastic face plate that screws on in place of the standard switch.

Project 9: An Inexpensive Burglar Alarm

Special switching mats can be bought that close contacts when stood upon. They are very thin and almost undetectable under mats and rugs. Also available from many sources are reed switches that can be fitted to windows and doors. They are operated by small permanent magnets concealed in the door. Whilst the door or window is shut the magnets are positioned close enough to the reed switch to keep it closed. On opening the door or window the switch opens.

These components, together with the circuit of Fig. 178, form the basis of a burglar alarm. The action of the circuit has already been explained in the chapter on photocells. The circuit of Fig. 190 is almost identical. If the potential of the base of the BC109 falls, the relay is energized. A fall in potential can be caused by any reed switch opening or any mat switch closing. It is necessary to have a latching arrangement so that the alarm bell continues to ring even if the original switch positions are regained. The bell can be stopped only by the householder. The reset switch can be a key-operated switch located in a concealed position.

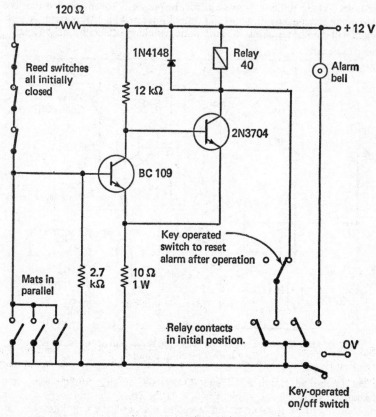

Fig. 190. Burglar alarm system. Key operated switches can be in remote secret locations.

Project 10: A Light-activated Switch

Fig. 191 shows the circuit of an arrangement that activates a relay when the light falling on the photocell is reduced to a low level; the circuit therefore is that of a dark-activated switch. Such a switch is useful in the home for switching on lights automatically when it becomes dark. The circuit may be used as a burglar deterrent since hall or lounge lights can be switched on when dusk approaches even when the occupant of the house is absent. A further example of its use is the automatic illumination of the area in front of garage doors at night. The driver is relieved from the chore of leaving the car and groping in the dark inside the garage for the light switch.

The location of the photocell is important, since it is undesirable to have the circuit revert to its original state when the cell is illuminated by the light from the bulb. Where difficulties arise in this respect it is recommended that the self-latching arrangement, shown in dotted lines in Fig. 191, is used. A reset switch will then be required; the automatic switching-off facility is then lost.

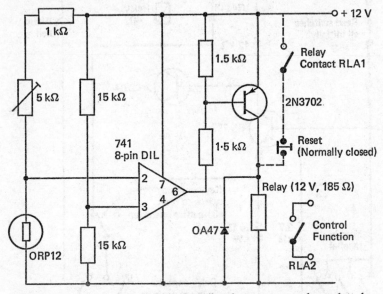

Fig. 191. Dark-activated switch. By including the components shown dotted, a self-latching facility is added.

The integrated circuit amplifier (741) is used in this configuration as a comparitor. The voltage at pin 3 is 6 V. If the voltage at pin 2 is a fraction less than 6 V then, since no negative feedback is used, the amplifier saturates and the voltage at pin 6 rises to almost 12 V. This occurs when the LDR is illuminated. Under this condition the resistance of an ORP 12 cell is low (130 Ω at a light level of 1000 lux; 2.4 kΩ at a level of 50 lux). The potential divider formed by the ORP 12 and variable resistor provides a low voltage at pin 2. When the cell is illuminated therefore the high voltage produced at pin 6 causes the 2N3702 (*pnp*) transistor to be cut off, and the relay is not energised.

When the cell is in darkness its resistance is several megohms. The potential at pin 2 therefore rises. Once the voltage at pin 2 exceeds 6 V the amplifier saturates in the opposite sense and the voltage at pin 6 is reduced to almost zero volts. The transistor is then turned on and the relay is activated.

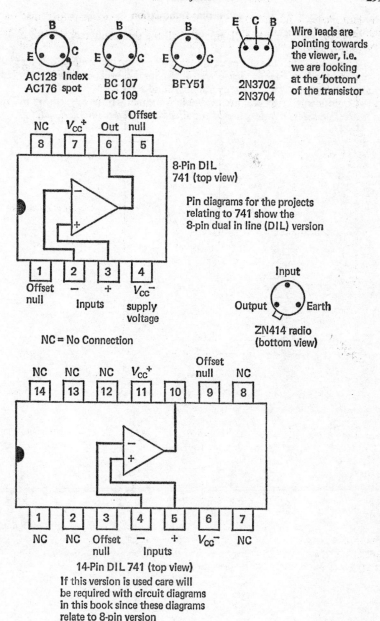

Fig. 192. Pin diagrams for semiconductors used in the projects described in this book.

Further Information

Fig. 192 shows the pin diagrams of all the semiconductor devices used in these projects.

Those who wish to start practical work at a more elementary level may care to consult *Electronics for Schools*, by R. A. Sparkes and published by Hutchinson Educational Ltd., 3 Fitzroy Square, London W1.

Components in general are available from many firms advertising in *Practical Electronics*, *Wireless World* and *Electronics Today International*.

INDEX

Index

Acceptor impurity, 45
Ampere, 9
Amplifiers
  audio, 72
  bandwidth, 86
  classes, 111
  Class A, 112
  Class AB, 113
  Class B, 111, 112
  d.c., 72
  design, 78
  emitter-coupled, 101
  feedback, 104
  frequency response, 81, 87
  high-fidelity, 249
  intermediate frequency (I.F.), 188
  long-tailed pair, 101
  power, 110, 290
  push–pull, 111, 117
  recording, 249
  tuned, 184
  Y-amplifier, 210
Amplitude, 7
Amplitude modulation, 175
Atomic structure, 1
Audio amplifiers, 72
Automatic gain control (A.G.C.), 190

Baffle, 255
Bandwidth, 86
  amplifier, 86
Base, 56, 57
Batteries, 121
BCD numbers, 271
Biasing circuit, 75
Bistable, 268
Boolean algebra, 262
Bridge rectifier, 132
Burglar alarm, 296
Burst-fire control, 149

Camera tube, 219
  colour tube, 229

Capacitors, 21
  ceramic, 26
  charging, 23, 37
  discharging, 24, 25, 37
  general-purpose, 26
  mica, 26
  reactance, 24
  variable, 28
Cathode-ray
  oscilloscope, 203
  tube, 203
Class A, 112
Class AB, 113
Class B, 111
Coercivity, 247
Collector, 56, 57
Colour code, 11
Colour television, 227
Colpitts oscillator, 157, 158
Common-emitter mode, 59
Common-mode rejection ratio, 102
Computer, digital, 274
Conductance, 10
Conductor, impurity, 45
Counting, 271
Crossover distortion, 115
Crossover networks, 257
Crystal microphone, 235
Crystal oscillator, 159

Damped oscillation, 153, 154
Darlington pair, 89, 90
Decibel, 82
Demodulator, 182
Detector
  A.M., 182
  F.M., 194
  Ratio, 198
Digital circuits, 262
Digital computers, 274
Digital integrated circuits, 98

Diodes
  Junction, 48, 50
  point contact, 51
  zener, 52
Disc records, 238
Disc store, 275
Discriminator, phase-shift, 195
Distortion, 83
  crossover, 115
  harmonic, 84
  intermodulation, 84
  non-linear, 84
  phase, 85
Dolby system, 252
Donor impurity, 45
Drain, 62, 63

Ear, characteristics, 232
Electromagnetic radiations, 173
Electromotive force, 6
Electron gun, 205
  lens, 205
Electrons, 1
Emitter, 56, 57
Emitter-coupled amplifier, 101
Emitter follower, 89

Feedback, 82, 85, 91
  amplifier, 104
Field effect transistor, 62
Flip-flop, 268
F.M. detector, 194
  radio, 256
  receiver, 200
Forward-bias, 50
Frame timebase, 223
Frequency modulation, 192
Frequency response, 81, 87
  amplifier, 81, 87
Full-wave rectifier, 132

Gain
  closed loop, 87
  open loop, 87
Gas sensor, 67
Gate, logic, 266
Gate, transistor, 62, 63
Germanium, 44

Half-wave rectifier, 48, 51, 131

Heatsink, 110
Heterodyning, 187
High-fidelity equipment, 234

Impedance, 88
Impurity conduction, 45
Indicating devices, 272
  hot filament, 271
  LEDs, 272
  liquid crystal, 273
  seven-segment, 272
Inductors, 29
  reactance, 31
Integrated circuit, 96
Interference, radio-frequency, 148
Ions, 1, 43

Junction diode, 48, 50

Light-dependent resistors (LDRs), 280
Linear integrated circuits, 98
Liquid-crystal indicators, 273
Lissajous figures, 213
Logic controller, 266
  gate, 266
Loudspeakers, 254

Magnetic store, 275
Matching, of load, 110
Microphones, 234
  capacitor, 237
  crystal, 235
  dynamic, 235
  moving-coil, 235
  ribbon, 236
Modulation
  amplitude, 175
  frequency, 192
Modulation index, 176
Modulator, cathode-ray tube, 205
MOS transistor, 64
Moving coil
  loudspeaker, 254
  pick-ups, 241
Multimeter, 294
Mutiple speaker system, 256
Mutual inductance, 32

n-type semiconductor, 45
Negative feedback, 82, 85

Oscillator, 152
  Colpitts, 157, 158
  crystal, 159
  Hartley, 157, 158
  LC, 155
  RC, 161
  relaxation, 167
  square-wave, 167
Oscilloscope, 207

p-type semiconductor, 46
Periodic time, 7, 34
Phase control of thyristors, 142
Phase shift discriminator, 195
Photocells, 278
  photoconductive (LDRs), 280
  photoemissive, 278
  photomultiplier, 279
  photovoltaic, 282
Photomultiplier, 279
Phototransistor, 284
Photovoltaic cells, 282
Pick-ups, 238
  ceramic, 240
  magnetic, 241
  stereo, 241
pn junction, 46
Post-deflection acceleration, 206
Potential difference, 4
Power amplifiers, 110, 290
Power supplies, 121
Preamplifier, 250
Push–pull amplifier, 111

Quadraphony, 242

Radio-frequency interference, 148
Radio waves, 173
Ratio detector, 198
RC oscillator, 161
Reactance, 24, 31
Receiver
  A. M., 181
  F.M., 192
  radio, 288, 289
  superhet, 187
  television, 220
Recording
  amplifier, 249

head, 247
tape, 245
Rectification, 48, 131
Rectifier
  bridge, 132
  full-wave, 132
  half-wave, 131
Relaxation oscillator, 167
Reservoir capacitor, 134
Resistors
  carbon film, 14
  cermet, 15
  colour code, 11
  fixed, 11
  high-stability, 14
  parallel, 15
  series, 15
  variable, 14
  voltage-sensitive, 21
  wirewound, 14
Reverse bias, 50
Rumble, 239

Schmitt trigger, 281
Semiconductor
  devices, 40
  junction, 40
Sensors
  gas, 67
  temperature, 66
Shadowmask tube, 230
Sidebands, 177
Silicon, 44
  n-type, 45
  p-type, 46
Sine waves, 33
Skip distance, 179
Sky waves, 178
Smoothing circuit, 133
Solar cell, 125
Source (FET), 62, 63
Square wave, 9, 167
Stabilization, voltage, 135
Stereo pick-up, 241
Superhet receiver, 187
Synchronization, 209, 222
Switch
  relay, 264
  transistor, 264

Tank circuit, 154
Tape recorder, 244
Television, 217
  colour, 227
  receiver, 220, 225
Temperature sensor, 66
Thermistors, 18
Thermometer, 19
Thyristor, 137
  control of a.c. power, 142
Time base, 207, 223
Time constant, 37
Tone controls, 253
Transducers, 2, 240
Transformer, mains, 126
Transistor, 41, 45
  field effect, 62
  junction, 55
  M.O.S., 64
  npn, 56, 57
  operation of, 57
  phototransistor, 284

pnp, 55
switches, 264
unijunction, 60
Transmitter, 180
Turntable, 238

Unijunction transistor, 60

Video detector, 226
Voltage, 4
  regulator, 136
  stabilizer, 135

Waveforms, 4, 35, 36, 37, 233
Wheatstone bridge, 16
Wien bridge oscillator, 164

Y-amplifier, 210

Zener diode, 52